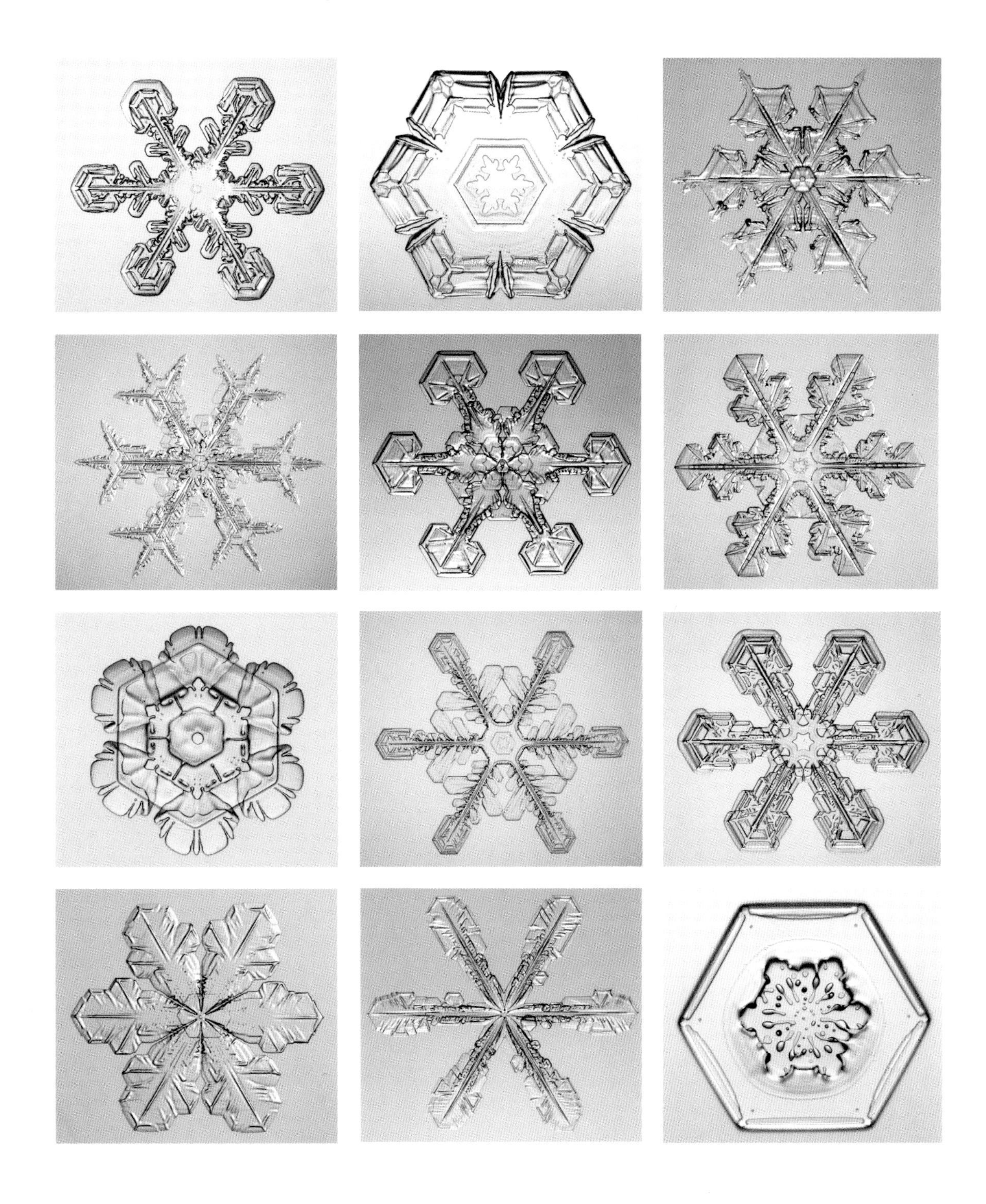

patagonia

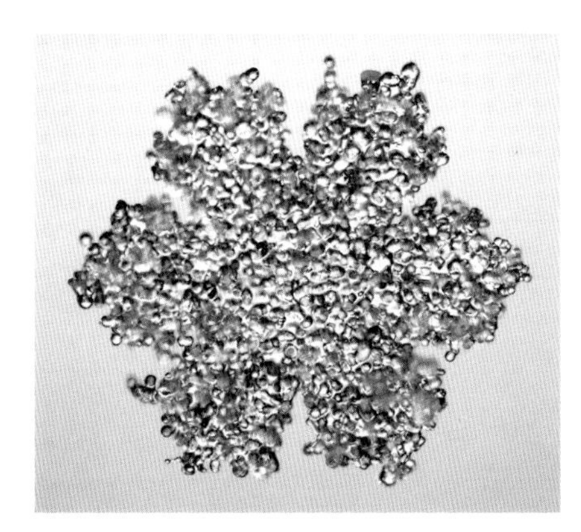
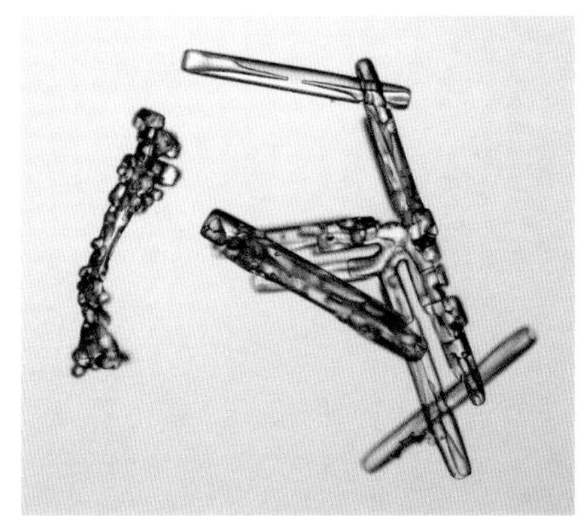

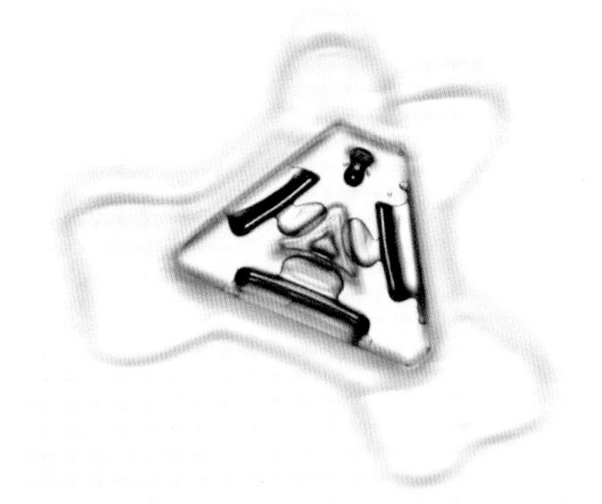

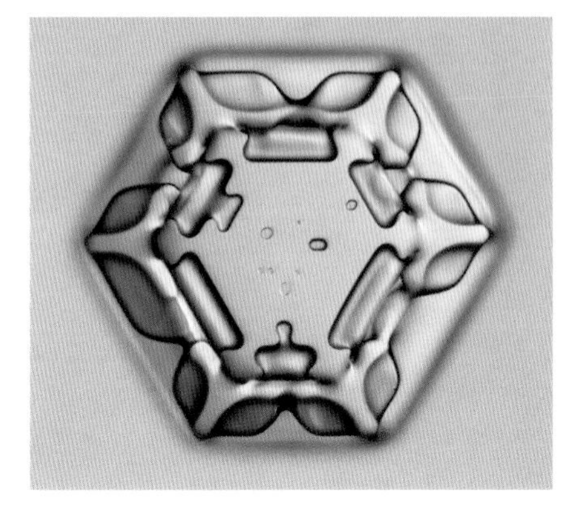

The Snowflake

Winter's Secret Beauty

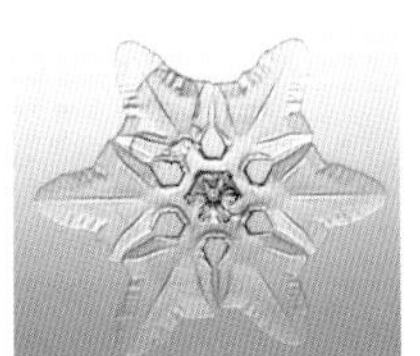 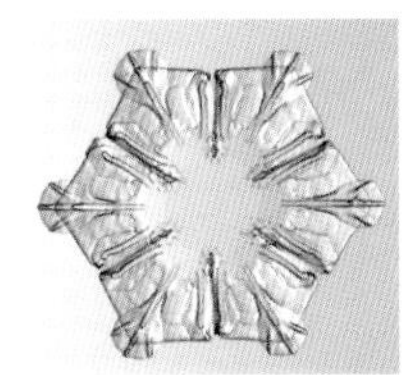 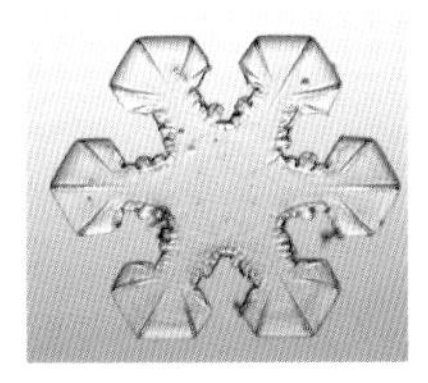

Text by Kenneth Libbrecht
Photography by Patricia Rasmussen

Colin Baxter Photography, Grantown-on-Spey, Scotland

First published in Great Britain in 2004 by
Colin Baxter Photography Ltd
Grantown-on-Spey
PH26 3NA
Scotland

First published in the United States by Voyageur Press, Inc.,
Stillwater, MN U.S.A.

A CIP Catalogue record for this book is available from the British Library

ISBN 1-84107-253-2

Printed in China

On page 1: Snow falls on a stand of hemlocks. (Photograph © Paul Rezendes)

On page 4: A young girl catches snowflakes on her tongue.
(Photograph © Layne Kennedy)

On page 5: Making snow angels in fresh-fallen snow.
(Photograph © Richard Hamilton Smith)

On pages 8 and 9: Snow blankets the Teton Mountains above the Snake River in Wyoming. (Photograph © Willard Clay)

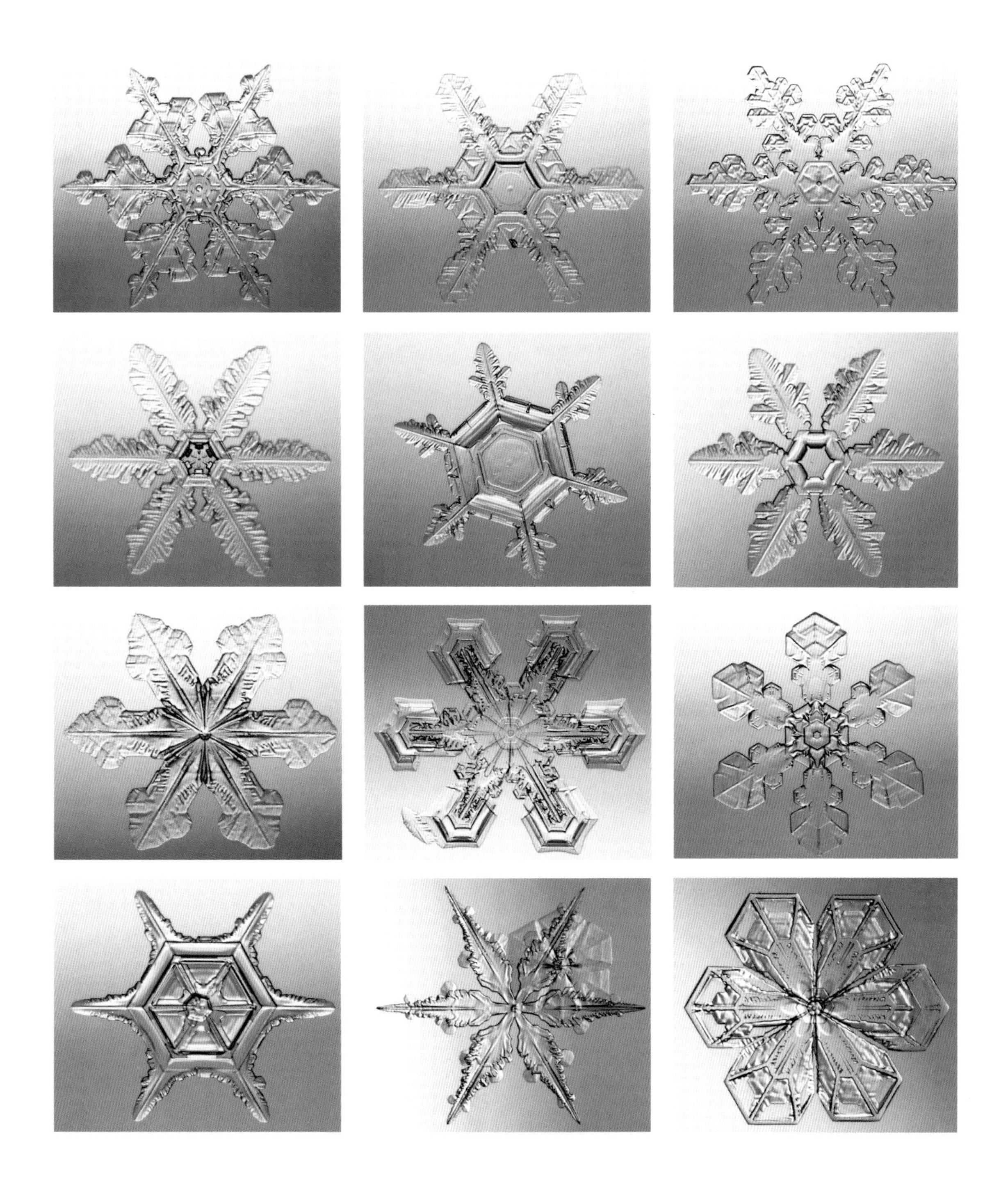

Contents

 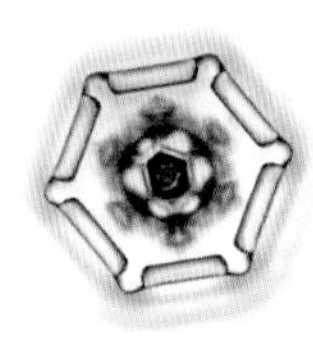 

Chapter 1 The Creative Genius 17

Chapter 2 Snowflake Watching 27

Chapter 3 Snow-Crystal Symmetry 35

Chapter 4 Hieroglyphs from the Sky 43

Chapter 5 Morphogenesis on Ice 51

Chapter 6 Snowflake Weather 59

Chapter 7 A Field Guide to Falling Snow 67

Chapter 8 In Search of Identical Snowflakes 101

Afterword Photographer's Statement *by Patricia Rasmussen* 107

Index 110

About the Author and Photographer 111

Chapter 1

THE CREATIVE GENIUS

"How full of the creative genius is the air in which these are generated!
I should hardly admire them more if real stars fell and lodged on my coat."
—Henry David Thoreau, Journal, *1856*

Growing up on a farm in North Dakota, I developed an early and intimate familiarity with snow. The cold north winds blowing down from the Canadian prairie brought us everything from quiet snow flurries to great howling blizzards. My winters were filled with snowballs, snow forts, and snowmen, with sliding, sledding, and skiing. Snow wasn't just part of the landscape where I lived; it was part of our heritage.

On snowy afternoons at school, our class sometimes trekked outdoors with magnifying glasses in hand to examine falling snowflakes. The crystals were especially well-formed on colder days, when the starlets would sparkle

COUNTLESS SNOWFLAKES
The simple pleasures of countless snowflakes on a winter's day—building a snowman and throwing snowballs. (Photograph © Richard Hamilton Smith)

brightly and linger long enough for a careful inspection of their shape and symmetry. Then the activity became a frenzied treasure hunt as we vied to see who could find the largest or most spectacular specimen.

Although I've admired many a snowflake in my day, only recently have I begun to appreciate their more subtle features. In my youth, the phenomenon of snow was so familiar that I usually ignored how extraordinary it is, that nature somehow manages to craft these miniature ice masterpieces right out of thin air. Perhaps we just had too much of a good thing—it can be difficult to appreciate the inner beauty of snowflakes when the driveway is piled high with them and you have a shovel in your hand.

It was only later in life, after moving to southern California and enjoying a long hiatus from shoveling, that I began to look carefully at snowflakes. I had been examining the physics of how crystals grow and form patterns, and I suppose my heritage returned to guide my thoughts. Snowflakes are growing ice crystals, after all. Before long I was studying just how these frozen structures are created in the clouds, and even how to engineer synthetic snowflakes in my laboratory.

Investigating the physics of snowflakes is an unusual occupation, to say the least. When visitors stop by my lab, they're sometimes puzzled why anyone would spend time trying to understand snowflakes. Was I trying to work out schemes for weather modification? Could this research improve the quality of artificial snow for skiing?

No, my flaky studies are not driven by practical applications. Instead my motivation is scientific curiosity—a desire to understand the material properties of ice and why it develops such elaborate patterns as it grows. The formation of snowflakes touches on some fundamental questions: How do crystals grow? Why do complex patterns arise spontaneously in simple physical systems? These basic phenomena are still not well understood.

Many materials form complex structures as they grow, and in the case of snowflakes, we see the results falling from the sky by the billions. I sought to understand how this works.

The Snowflake Menagerie

A snowflake is a temporary work of art. To capture the images in this book, each snowflake was plucked from the air as it fell and then rapidly photographed. In mere minutes, sometimes seconds, a fallen snowflake starts to lose its shape. The sharp corners begin to round, and after a brief time the delicate features are gone. No two snowflakes look exactly alike when they fall, but their uniqueness is soon lost on the ground. Once inside a snowbank, intricately patterned snowflakes quickly transform into tiny, formless lumps of ice.

When I say *snowflake*, I usually mean *snow crystal*. The distinction between the two derives from their meteorological definitions. A *snow crystal*, as the name implies, refers to a single crystal of ice. A *snowflake* is a more general term that can mean an individual snow crystal, a cluster of snow crystals that form together, or even a large aggregate of snow crystals that collide and stick in midair, falling to earth in a flimsy puffball. Snow crystals are commonly called snowflakes, which is fine, like calling a tulip a flower.

When most people think of snowflakes, they think of elaborate, multi-branched snow stars. These are the ever-popular icons of ski sweaters and winter-holiday decorations. Nature produces a great many variations of this type of snow crystal, each exhibiting its own style of branching and sidebranching. Some stellar crystals contain scores of sidebranches, giving them a leafy, almost fern-like appearance. Others contain fewer sidebranches, perhaps decorated with thin, patterned ice plates.

Whatever their appearance, stellar snow crystals usually grow six primary branches, each supporting additional sidebranches. Sometimes the side-branching appears to be symmetrical, but often it does not. Occasionally you can even find snow stars with three- or twelve-fold symmetry.

One thing you will not find in nature is an eight-sided snow crystal. The same is true of four-, five-, and seven-sided snow crystals. The symmetry of the ice crystal does not allow such forms. Eight-sided snowflakes

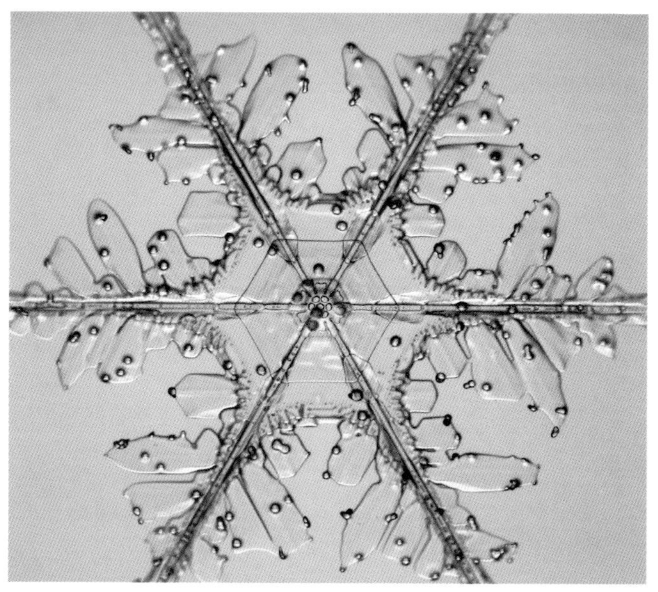

Stellar Snowflakes

Snow crystals often fall to earth as six-branched ice stars, created in endlessly different patterns.

Cloud Droplets

Left: This snow crystal features small ice particles called *rime*, water droplets that struck the crystal and froze on its surface as it fell through the clouds.

SNOWFLAKE CLUSTERS
These three large stellar crystals formed a small snowflake cluster when they collided in midair and stuck together.

may be easier to cut out of paper, but real snow crystals never have eight-fold symmetry, regardless of what you see in holiday decorations.

Another common, but lesser known, snow-crystal type consists of long, columnar or needle-like crystals. The most basic form these can take is a simple hexagonal column of ice, similar to the shape of a standard wooden pencil. Often the columns are hollow and almost sheath-like. Other times they grow into assemblies of thin ice needles. A snowstorm will occasionally bring great numbers of these sharp shards of ice, and they're something to be reckoned with when blown into your face by a stout wind.

The snowflake menagerie includes many variations on these simple themes—plates, columns, branched, sectored, hollowed, six- and twelve-sided. There are all sorts of ornate and curious designs that can be seen, all floating lazily to earth in vast numbers. One does not easily become bored looking at snowflakes.

Not just any snow-crystal shape or pattern can be seen falling from the sky, however. The patterns that appear are not random, but are determined by the rules that govern snow-crystal formation. Just as the colors of the stars reveal their composition, the shapes of snow crystals tell a tale of how they were created.

STELLAR DENDRITES
Stellar crystals often appear in fern-like *dendrite* forms that contain numerous sidebranches. This one also picked up a few broken branches from other crystals. Stellar dendrites are the largest of the snow-crystal menagerie. This one is nearly 5 mm (0.2 inches) from tip to tip, about the size of a small pea.

The Making of a Snowflake

Snowflakes are made of ice, yet ice alone does not a snowflake make. You could produce a million ice cubes in your freezer and not one would look even remotely like a beautiful stellar snow crystal. You cannot simply freeze water to make a snowflake; you have to freeze water in just the right way.

The mystery of snowflakes is how they are fashioned into such complex and symmetrical shapes. Snowflakes are not made by machines, nor are they alive. There is no blueprint or genetic code that guides their construction. Snowflakes are simple bits of frozen water, flecks of ice that tumble down from the clouds. So how do they develop into such intricate six-branched structures? Where is the creative genius that designs the neverending variety of snow-crystal patterns?

Many people think snowflakes are made from frozen raindrops, but this is simply not true. Raindrops do sometimes freeze in midair as they fall, and this type of precipitation is called *sleet*. Sleet particles look like what they are—little drops of frozen water without any of the ornate patterning or symmetry seen in snowflakes.

You do not make a snowflake by freezing liquid water at all. A snowflake forms when water *vapor* in the air condenses directly into solid ice. As more vapor condenses onto a nascent snow crystal, the crystal grows and develops, and this is when its elaborate patterning emerges. To explain the mystery of snowflakes, we must look at how they grow.

In a snowflake, just an ordinary snowflake, we can find a fascinating story of the spontaneous creation of pattern and form. From nothing more than the simple act of water condensing into ice, these amazing crystal structures appear—complex, symmetric, and in endlessly varying designs. Snowflakes are the product of a rich synthesis of physics, mathematics, and chemistry. They're even fun to catch on your tongue.

Snowflake Size

A typical snow crystal measures about 1-2 mm (0.04-0.08 inches) in diameter, just large enough to fit inside the letter "O" printed here. These ornate slivers of ice are large enough to identify on your coat sleeve with an unaided eye but small enough that special equipment is needed to reveal their inner structure.

Asymmetrical Snowflakes

Above: Most snow crystals are not beautifully symmetric, but are more irregular in shape. Some, like this one, are downright odd.

Hollow Columns

Left: Columnar snow crystals may be less beautiful than their stellar cousins but are more common. This form can be found in abundance during warmer snowfalls. Many columns, like this one, include two conical-shaped hollow regions inside the crystal.

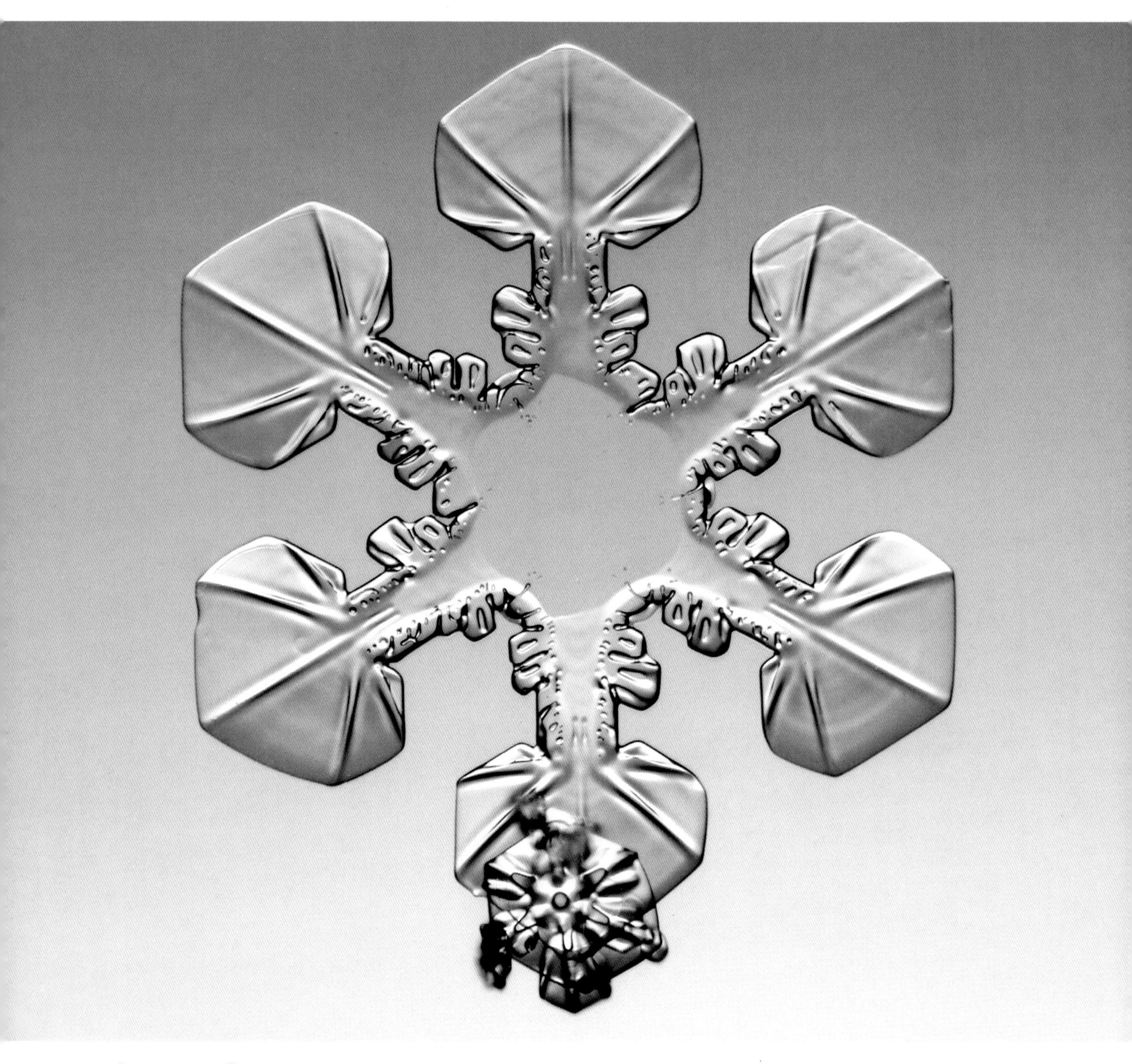

SECTORED PLATES

Snow crystals often develop thin ice-plates on the ends of their arms. Each plate here features ridges dividing it into sectors; thus, these are called *sectored plates*. You can find ridges like these hiding in many snow crystals if you look closely.

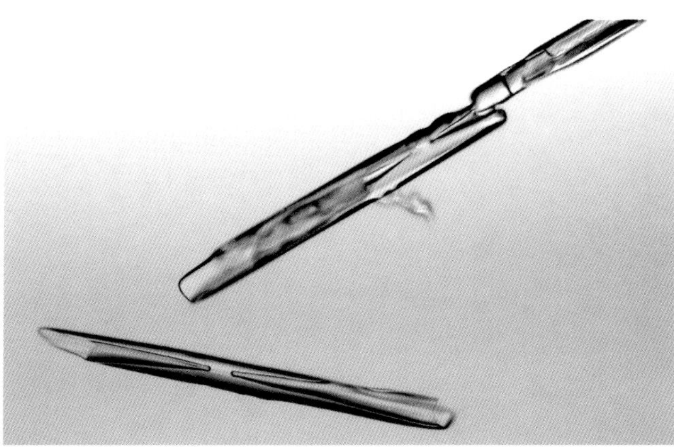

Twelve-Sided Snowflake

Above: Twelve-sided snowflakes are uncommon but worth the search. This one is essentially two six-branched snow crystals joined in the middle, each growing independently of the other.

Needle Crystals

Left: Some columnar crystals grow into slender, needle-like snow crystals that can sting when blown into your face by a wind. The lower crystal is a thin hollow column with sharp ends, 1.6 mm (0.06 inches) in length.

Chapter 2

SNOWFLAKE WATCHING

 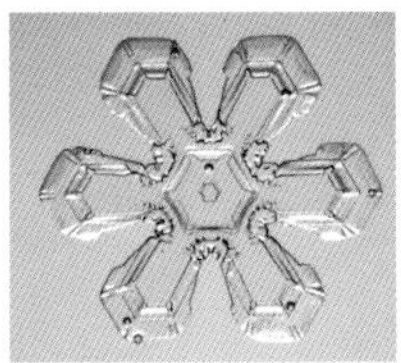 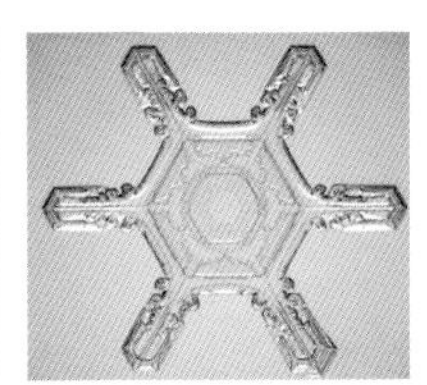

"No weary journeys need be taken, no expensive machinery employed. . . . A winter's storm, an open window, a bit of fur or velvet, and a common magnifier, will bring any curious inquirer upon his field of observation with all the necessary apparatus, and he has only to open his eyes to find the grand and beautiful laboratory of nature open to his inspection."

—*Frances Chickering,* Cloud Crystals: A Snow-Flake Album, *1864*

The earliest detailed account of snow-crystal structure was penned by French philosopher and mathematician René Descartes in 1637. Renowned for his pioneering work uniting the then-disparate mathematical subjects of geometry and algebra, Descartes is perhaps best remembered for his metaphysical dictum *Cogito, ergo sum* (I think, therefore I am). In his scientific study of meteorology and weather, *Les Météores*, Descartes recorded thorough naked-eye observations of snow crystals, including several rarer forms:

CLOUD CRYSTAL

Facing page: A singular "cloud crystal" from "the grand and beautiful laboratory of nature."

"After this storm cloud, there came another, which produced only little roses or wheels with six rounded semicircular teeth . . . which were quite transparent and quite flat . . . and formed as perfectly and symmetrically as one could possibly imagine. There followed, after this, a further quantity of such wheels joined two by two by an axle, or rather, since at the beginning these axles were quite thick, one could as well have described them as little crystal columns, decorated at each end with a six-petalled rose a little larger than their base. But after that there fell more delicate ones, and often the roses or stars at their ends were unequal. But then there fell shorter and progressively shorter ones until finally these stars completely joined, and fell as double stars with twelve points or rays, rather long and perfectly symmetrical, in some all equal, in others alternately unequal."

Descartes was obviously impressed with the geometrical perfection of snowflake forms. The use of mathematics to describe the broad spectrum of ordinary phenomena was still a new idea at the time, and a major step forward in the development of the natural sciences.

The invention of the microscope in the mid-seventeenth century quickly led to more and better snowflake observations. English scientist and early microscopist Robert Hooke sketched snowflakes, along with practically everything else he could find, for his famous book *Micrographia*, published in 1665. Although his crude microscope was hardly better than a good magnifying glass today, Hooke's drawings nevertheless began to reveal the complexity and intricate symmetry of snow-crystal structure, details that could not be detected with the unaided eye.

As the quality and availability of optical magnifiers improved, so did the accuracy of snow-crystal drawings. In the mid-nineteenth century, a number of scientists and amateurs around the world recorded the diverse character of snow-crystal forms. Snowflakes are quite short-lived, however, so snowflake artists inevitably relied on memory to complete their sketches. As a result, even the best snow-crystal drawings lacked detail and were not completely faithful to their original subjects.

One of the first books devoted entirely to snowflakes appeared in 1864 with the publication of *Cloud Crystals: A Snow-Flake Album*. The author was snowflake artist Frances Knowlton Chickering of Portland, Maine. Little is known about Mrs. Chickering beyond that she was a minister's wife and a diligent observer of natural phenomena, with a keen eye for snowflakes. During Maine's long winter months, she examined falling crystals and from memory quickly cut paper outlines of their forms. These images represent some of the earliest artistic renditions of snow crystals and include a number of lesser-known snow-crystal forms—fern-like dendritic stars, sectored plates, twelve-sided snowflakes, and capped columnar structures were all recorded.

With just a simple magnifying glass, Mrs. Chickering observed what fell from the sky with a clear mind and a propensity for thorough scrutiny. To use her own words, she opened her eyes to find the grand and beautiful laboratory of nature before her. Mrs. Chickering was a true pioneer in the art of snowflake watching.

EARLY SNOWFLAKE SKETCHES

These crude drawings were published in 1665 in *Micrographia* by English scientist Robert Hooke, who made the first observations of snow crystals using a microscope. (Huntington Library, San Marino, California)

Arctic Snow Crystals

English explorer William Scoresby made these sketches during a winter voyage through the Arctic, which he recounted in his 1820 book *An Account of the Arctic Regions with a History and Description of the Northern Whale-Fishery*. They are the first drawings that accurately depicted many details of snow-crystal structure, as well as several rare forms. Scoresby noted that the cold arctic climate produced more highly symmetrical crystals than typically seen in Britain.

Snowflake Cutouts

Frances Chickering, a minister's wife from Maine, published these snow-crystal images in her 1864 book *Cloud Crystals: A Snow-Flake Album*. She examined snow crystals as they fell on her windowsill and quickly cut paper outlines of their forms.

CAPTURING SNOWFLAKES
Vermont farmer Wilson Bentley first developed the art of snow-crystal photography and produced a large album of snowflake images. He is shown here with his original snow-crystal photo-microscope. (Jericho Historical Society)

Snowflake Photography

It took Wilson Bentley, a farmer from Jericho, Vermont, to create the first photographic album of falling snow, thus awakening the world to the hidden wonders of snowflakes. Bentley became interested in the microscopic structure of snow crystals as a teenager in the late 1800s, and he soon began experimenting with the new medium of photography as a means of recording what he observed. He constructed an ingenious mechanism for attaching a camera to his microscope for this purpose, and succeeded in photographing his first snow crystal in 1885.

To say Bentley was dedicated to the task was an understatement. Snowflake photography became his lifelong passion, and in the course of forty-six years he captured more than 5,000 snow-crystal images on the old-style glass photographic plates.

Bentley's photographs appeared in countless publications, providing for many people their first look at the inner structure and symmetry and the incredible variety of snow crystals. And with image after image, Bentley's photos showed the world that no two snowflakes are exactly alike.

In the late 1920s Bentley worked with W. J. Humphreys, chief physicist for the United States Weather Bureau, to publish a book containing more than 2,000 snow-crystal images. *Snow Crystals* appeared in November 1931, and Bentley died of pneumonia just a few weeks later at the age of sixty-six. He resided his entire life in the same farmhouse, and for all his photographs he used the same photomicroscopy equipment he constructed as a teenager.

EARLY SNOWFLAKE PHOTOGRAPHS
These are but a few of the thousands of snow-crystal photographs taken by Wilson Bentley between 1885 and 1931. Bentley illuminated his crystals using daylight, but soon became dissatisfied with photos of bright snow crystals against white backgrounds. To alter them, he made duplicate negatives of each image and then painstakingly scraped the emulsion from around the crystal, a process that took about an hour per picture. His prints from these "blocked" negatives yielded bright snow crystals on black backgrounds. Bentley was always quick to point out that he never changed the snow-crystal images themselves. (Jericho Historical Society)

TREASURE HUNTING
Snowflakes come in a dazzling assortment of shapes and sizes, often with elaborate patterns and sometimes with a stunning simplicity.

Snowflake Watching

Imagine the billions upon billions of exquisite ice sculptures floating down from the sky during a single snowfall—all impermanent, all unnoticed. Each snowfall provides a unique exhibit of winter art. This show deserves an audience.

Snowflake watching is akin to birdwatching. Both are easy outdoor recreations focused on observing nature. Each requires only a minor investment in equipment. To observe snowflakes you need little more than an inexpensive magnifying glass; nearly any kind will do. Add to that a gentle snowfall, and these miniature masterpieces of nature can be witnessed firsthand. If the climate is agreeable, you too might be struck by the sight of a magnificent snow star, or your first glimpse of a twelve-sided snowflake. You never know what you might discover.

On cold days, I find that a parked car makes an excellent snowflake observatory. The windshield is smooth and about the right height, the slope of the glass makes for comfortable viewing, and old snowflakes can easily be brushed aside to make room for freshly fallen ones. When the weather is warmer, a piece of black cardboard or a sleeve of insulating fleece works well.

Whatever collecting surface you prefer, cold is a must. If the surface is warmer than the ambient air, captured snowflakes will quickly degrade. Body heat from hands or a misplaced breath can also have detrimental effects. It's not enough to protect yourself from the cold; when snowflake watching, you must protect the cold from you.

One thing you quickly discover as a snowflake watcher is that each snowfall has its own personality. Some storms bring great numbers of small, plate-like crystals. Others produce only columnar forms. The character of the crystals often changes dramatically during the course of a single snowfall.

At times, especially when it's relatively warm, a storm will bring little more than gloppy clumps that offer little for study. The crystals melt a bit in flight, or they collide and stick together before hitting the ground. As Henry David Thoreau observed in his *Journal,* "commonly the flakes reach us travel-worn and agglomerated, comparatively without order or beauty, far down in their fall, like men in their advanced age."

On occasion, however, a snowfall will bring quantities of spectacular stellar crystals that fall lazily on your sleeve, minute but magnificent ice flowers. Every snowfall brings something different, and every snowflake tells its own story.

In snowflake watching, like birdwatching, your quarry may not always appear when you want it to, and you shouldn't expect to see extraordinary specimens your first time out. The best strategy is to be prepared by carrying a magnifying glass whenever conditions are favorable. It need not be the large, Sherlock Holmes variety; a small, fold-up pocket magnifier, tucked inside a coat pocket, will do splendidly.

On a day when the crystals are falling well you will be amply rewarded for your efforts. You may even attract a crowd of spectators, on the ski slope or wherever, waiting for a turn to use your lens. Although a simple magnifying lens cannot reveal all the minute details seen through a microscope, there are many remarkable sights you may find on your coat sleeve. Photographs cannot convey the sparkle you see as you move a crystal around to observe the light play on its facets. Add to that the quiet pleasure of watching crystals float lazily to earth, plus the excitement of discovering new forms, and you have a winter recreation that is quite unlike anything else.

Snow-Crystal Hot Spots

Where do you go if you're a determined snowflake watcher and you want to see some really great snowflakes? Are mountain snowflakes different from those that fall at low elevations? Will you find rare snow-crystal types in arctic regions that cannot be found at lower latitudes?

No one really knows the answers to these questions. While we have decades of snowfall records around the world, there are almost no meteorological records of what *types* of snowflakes fell when and where. Places that experience the most snowfall do not necessarily produce the most interesting snowflakes.

There are several sites around the globe that have gained snow-crystal notoriety over the years, mainly through the work of a few individuals like Wilson Bentley who have looked carefully at falling snow. In North America, areas around the Great Lakes are known to produce fine snow crystals; the images in this book were captured in Wisconsin, for example. Locations in New England are recognized for the quality and quantity of their snow, and I've found North Dakota snowflakes can be pretty impressive as well. Northern Japan, particularly central Hokkaido, is a prime spot for observing large stellar crystals. We might also expect to find great snow crystals across Canada, Alaska, Greenland, northern Asia, Scandinavia, Antarctica, and perhaps in many mountainous regions around the world, but we have little solid data one way or the other.

There is an old saying among geologists and miners that gold is where you find it. Likewise, it's impossible to predict where the best snowflake-watching spots will be—good snow crystals are where you find them. Some areas are well suited for producing ornate stellar crystals. Perhaps other spots are better for finding rare and exotic specimens. If you live in a cold climate, your backyard may be a terrific location for snowflake watching. All you need is a magnifying glass to find out.

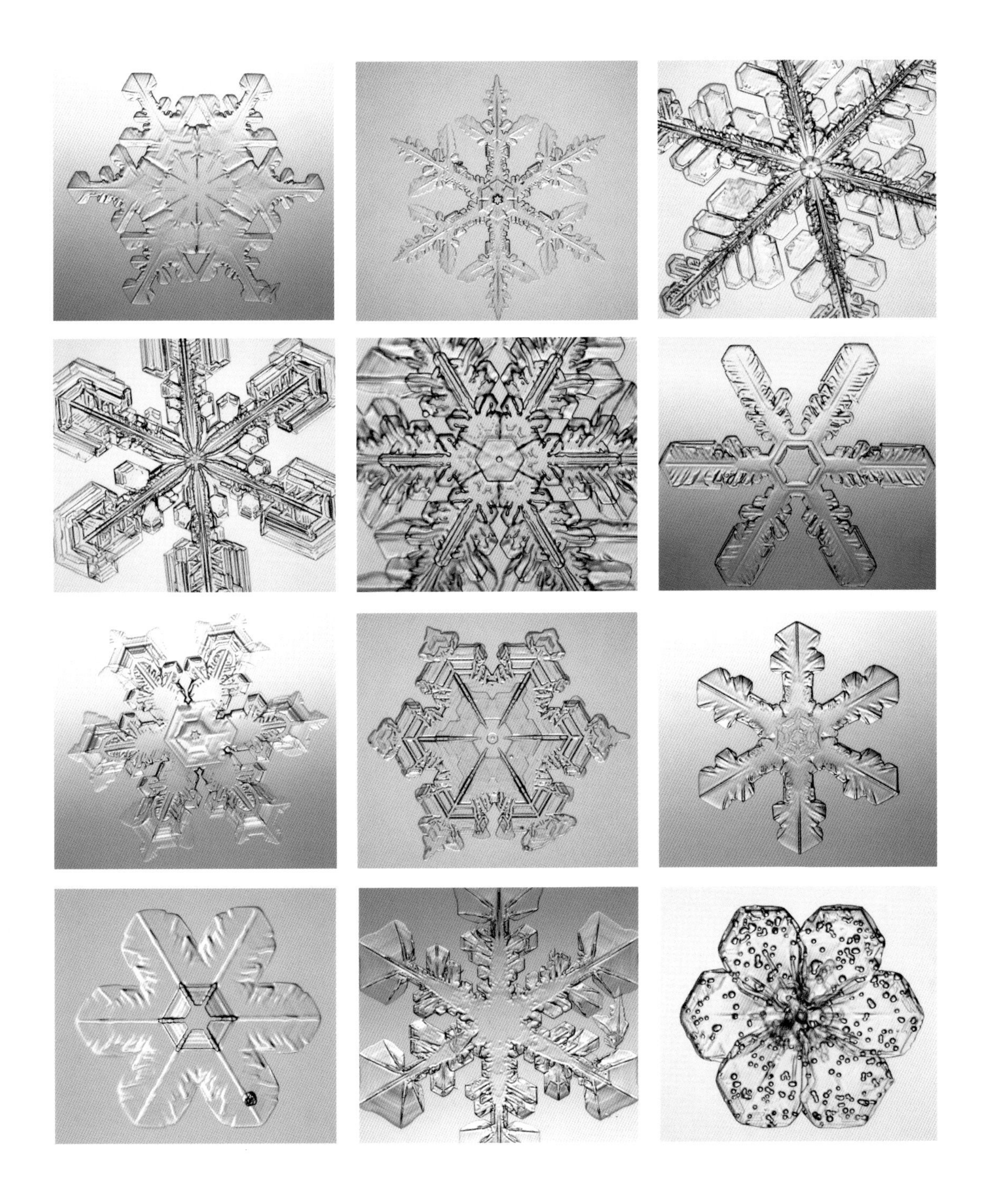

Chapter 3

Snow Crystal Symmetry

"Poets say science takes away from the beauty of the stars—mere globs of gas atoms. I too can see the stars on a desert night, and feel them. But do I see less or more? . . . What is the pattern, or the meaning, or the why? It does not do harm to the mystery to know a little about it. For far more marvelous is the truth than any artists of the past imagined it."

—Richard P. Feynman, The Feynman Lectures on Physics, *1963*

The first person to look at snow crystals with a scientific eye and scrutinize their remarkable six-fold symmetry was German scientist Johannes Kepler in the early seventeenth century. One of the most brilliant and influential scientists of his day, Kepler made numerous contributions to astronomy, mathematics, and optics, including his famous discovery of what are now called Kepler's Laws, describing the motions of the planets around the sun. Kepler also noted the precise six-fold symmetry of snow crystals, and sought to understand its origin.

In 1611 Kepler presented to his patron Emperor Rudolf II a small treatise entitled *The Six-Cornered Snowflake*. In his treatise, Kepler compared the

SYMMETRY

Facing page: Snow-crystal symmetry comes from nothing but the simple interactions between water molecules.

SIX-FOLD SYMMETRY

Like snow crystals, certain flowers such as these tulips also exhibit six-fold symmetry. The origin of floral symmetry, however, is buried deep within the genetic code and emerges due to complex biochemistry. (Photograph © Richard Hamilton Smith)

six-fold symmetry of snowflakes to similar symmetries found in flowers. He deduced that the similarities must be in appearance only, since flowers are alive and snowflakes are not:

"Each single plant has a single animating principle of its own, since each instance of a plant exists separately, and there is no cause to wonder that each should be equipped with its own peculiar shape. But to imagine an individual soul for each and any starlet of snow is utterly absurd, and therefore the shapes of snowflakes are by no means to be deduced from the operation of soul in the same way as with plants."

Kepler realized that living things were far beyond his comprehension—to the point that he ascribed to each flower an individual soul. Science has come a long way since Kepler's day, but the symmetry of living things remains enigmatic. The arrangement of petals on a flower is just one minor trait that derives from life's amazingly complex biochemical machinery. I daresay we will not comprehend all the biological details of a flower any time soon.

Kepler surmised that a snowflake was a simple thing without a soul. It was just a sliver of ice, after all. Therefore he believed it fruitful to question what organizing principle was responsible for its symmetry. He noted that cannonballs display a hexagonal pattern when stacked in a pile, and he conjectured that these two symmetries were related. There is a germ of truth in this reasoning, since the geometry of stacking atoms is at the heart of snow crystal symmetry. But the atomistic view of matter had not been developed by 1611, so Kepler could not carry the cannonball analogy far.

Kepler realized that the genesis of crystalline symmetry was a worthy scientific question. He also recognized the similarity between snow crystals and mineral crystals, both exhibiting symmetrical, faceted structures. But at the end of his treatise, Kepler accepted that the science of his day was not advanced enough to explain any of it. A good scientist knows when to admit ignorance and move on.

It would be 300 years before X-ray crystallography allowed physicists to observe the regular stacking of molecules in crystals, including ice, thus finally explaining crystalline symmetry. It is the crystal structure of ice—the regular arrangement of its water molecules—that is the underlying source of snowflake symmetry.

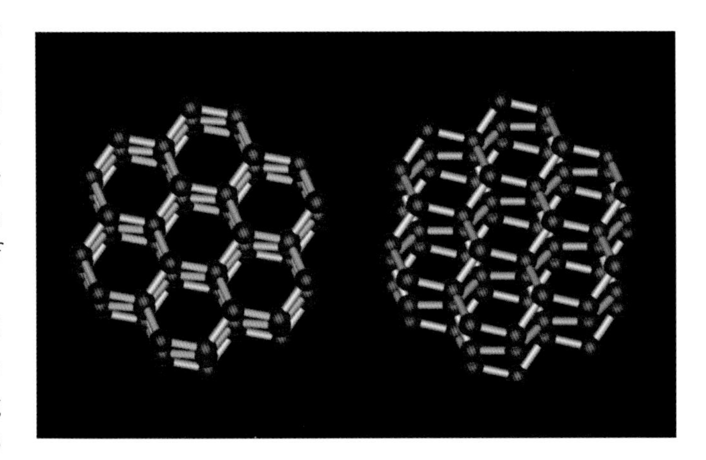

HEXAGONAL STRUCTURE

These two different views show the crystal structure of the normal form of ice, known as *Ice Ih*. The red balls represent oxygen atoms in the H_2O molecules; each gray bar is a hydrogen atom. The hexagonal lattice structure of the ice crystal is what ultimately gives snow crystals their six-fold symmetry.

Complex Symmetry

Many snow crystals show remarkably complex symmetry. This sliver of ice contains a great many small markings—and most are faithfully reproduced on each of the six arms. In addition to its six-fold symmetry, each arm of the crystal is also symmetric about its central axis.

Crystals

The word *crystal* derives from the Greek *krystallos*, which means "ice" or "clear ice." In spite of its meaning, *krystallos* was not originally used to describe ice, but rather the mineral quartz. The early Roman naturalist Pliny the Elder described clear quartz *krystallos* as a form of ice, frozen so hard that it could not melt. He was of course mistaken on this point: quartz is not a form of ice, nor is it even made of water. But amazingly enough, after nearly 2,000 years, Pliny's mistake is still felt in the language of the present day. If you look in your dictionary, you're likely to find one of the first definitions for crystal is simply "quartz." I object to my dictionary on this point; it's like saying the definition of food is "potato"! I blame Pliny.

The scientific definition of a crystal is any material in which the atoms or molecules are lined up in a regular array. Ice is a crystal made of water molecules, and the normal form of ice is called *Ice Ih*, made of sheets of water molecules arranged into "puckered" hexagons. Hexagons, of course, have six-fold symmetry, and this symmetry ultimately carries over into snow crystals.

Besides ice, all sorts of crystals can be found in our everyday lives. Copper is crystalline, as are ruby and diamond. Computer chips are made from silicon crystals. Most rocks are made from jumbled bits of crystalline minerals like quartz. Salt, sugar, and aluminum foil are a few crystalline materials you can pick up at your grocery store.

All crystals demonstrate an amazing organizational ability—they assemble themselves. A crystal's order and symmetry arise spontaneously, starting with a random collection of molecules. This organizational feat should not be overlooked. If you want a brick wall somewhere, it certainly does not assemble itself. You have to build it, brick by brick. The order and symmetry are imposed by you, the builder. It's absurd to think that the bricks could magically put themselves together into a wall.

Self-assembly is how things are made in the natural world—crystals, snowflakes, plants, animals. Even you and I are made from self-assembled parts, guided by biochemical rules. Yet self-assembly is hard to fathom because it usually involves either nanoscale objects, like the molecules in a crystal, or tremendously complex objects, like living things. We almost never get to see self-assembly happening directly.

Even so, here is one simple demonstration of self-assembly you can try in your kitchen. Take a bunch of toothpicks, dump them into a flat-bottomed bowl, and add a thin layer of soapy water. Now swish the mixture around a bit, and *voilà!*—the toothpicks will align themselves together into a clump. You've made a self-assembled toothpick "crystal."

It's not all that impressive at first glance, I suppose. The soapy water makes the toothpicks stick together, and they especially like to stick together side by side. Once they're stuck together, they tend to stay stuck. So after swishing the initially disordered mixture around a bit, you're left with an ordered arrangement of toothpicks, an order that arose spontaneously.

Ice crystals work in much the same way. Water molecules form chemical bonds between themselves, and these bonds make the molecules line up and stick together. The bonds have certain preferred orientations, and this dictates how the water molecules stack up. Thermal agitation jostles them into position, and soon you're left with an ordered arrangement of water molecules—an ice crystal.

ROCK CANDY AND ROCKS
Sugar, left, and the mineral quartz, right, are two crystals that often grow into faceted forms. (Photo © Kenneth Libbrecht)

Toothpicks and water molecules perform simple acts of self-assembly, but this is just the beginning of what nature can do with this tool. Atoms not only self-assemble into crystals, but also into small molecules. Small molecules self-assemble into larger molecules. Large molecules self-assemble into sheets, ribbons, helices, and membranes. Add a few more steps of self-assembly—make that quite a few more steps—and out pop proteins, chromosomes, ribosomes, and cell walls. Put all the self-assembled pieces together—lots more self-assembly steps—and you have a bacterium. No one knows how all this works in detail, but it obviously works well. Nature follows the simple physical and chemical rules of self-assembly to make everything, including snowflakes.

Ice Types

In the crystal world, ice is an inspired material. At last count, there are fourteen different stable or nearly stable types of ice, more than for any other known solid. Each represents a different way water molecules can be stacked into a solid. All except Ice Ih exist only under exotic conditions: extremely low temperatures, high pressures, or both. Most are formed at thousands of atmospheres of pressure, where water molecules are crushed into denser arrangements.

These high-pressure forms of ice are mostly things that only a physicist could love. Nevertheless, all good science can be turned into good science fiction, and even high-pressure ice had its day in the sun. Back when only seven forms of ice were known, author Kurt Vonnegut Jr. brought some lasting fame to the yet-undiscovered Ice IX, or ice-nine. Vonnegut wove an entertaining tale in his 1963 novel *Cat's Cradle*, in which ice-nine was an ultra-stable form of ice, capable of crystallizing all the world's water at room temperature. This was Pliny's *krsytallos* all over again—ice-nine was ice frozen so hard that it did not melt.

Of course, the real ice-nine was rather more benign when it was finally discovered, and all the exotic forms of ice revert to ordinary Ice Ih when they are brought to familiar pressures and temperatures.

The Little Faces

Snow crystals often display numerous reflecting facets, similar to those on a cut diamond. *Facet* literally means "little face," and the sparkles you see when the sun shines on a bank of snow are reflections from the countless little faces of snow crystals. You can also see the almost mirror-like surfaces of individual snow crystals when you look at them on your sleeve. As you move a crystal to and fro, the bright reflections quickly reveal the smooth facet surfaces.

This aspect of the precision construction of snow crystals was first recorded by Descartes, who described snow crystals in *Les Météores* as "little plates of ice, very flat, very polished, very transparent, about the thickness of a sheet of rather thick paper . . . but so perfectly formed in hexagons, and of which the six sides were so straight, and the six angles so equal, that it is impossible for men to make anything so exact."

Descartes was describing snow-crystal facets, and clearly these are special surfaces when it comes to the formation of snow crystals. But where do facets come from? Why are they flat, and how do they form?

Facets come in two varieties, natural and artificial. Again I must object to my dictionary, which only recognizes the latter by defining a facet to be "one of the small, polished plane surfaces of a cut gem." That definition may work well at the jewelry store, but who's up there polishing plane surfaces on all the snow crystals?

The facets in small gemstones are almost invariably artificial, regardless of the type of stone. The diamonds, emeralds, and rubies you find in jewelry all sparkle from their artificial facets, which were carved and polished with a grindstone.

Facets are also carved or molded into glass products, like ornate bowls and goblets. In yet another linguistic offense, these latter items are called fine "crystal" at the department store, although in fact the material used is glass. I find this misnomer particularly amusing because the molecules in glass are completely disordered, randomly jumbled atop one another. At the molecular level, a glass "crystal" is the very opposite of a crystal!

In contrast to artificial facets, natural facets arise from the self-assembly of crystalline materials. Quartz crystals, such as the purple-hued amethyst, can be pulled out of the earth in beautiful faceted shapes, as can garnets and many other gem-like minerals. As with snowflakes, the faceted appearance of these crystals arises spontaneously as the crystals grow.

Most crystals can grow into faceted shapes, but many only do so when they grow slowly. Metals like copper and aluminum are typically aggregates of extremely small crystals, so small that you cannot see any faceting without a microscope, and you would hardly know from looking at them that these materials are crystalline at all. Sugar is not always faceted, but with a little coaxing on your kitchen countertop you can grow some beautiful faceted forms. These are called "rock candy" crystals because of their resemblance to faceted minerals.

Naturally faceted crystals always show a characteristic symmetry in the angles between the facets. This symmetry comes from the chemical forces that determine the angles of the molecular bonds inside the crystal. The natural facets we see in crystals, however, including snowflakes, are far larger than the molecules inside the crystal. So the question arises: How can molecular forces, operating only at the nanoscale, determine the shapes of large crystals? How does one end of a crystal facet manage to grow the same as the other end? Does one end know what the other end is doing?

The molecules on opposite corners of a growing faceted crystal do not communicate with one another to determine the crystal shape. Nor do they have to. The reason facets form is simply because some surfaces acquire material and advance more slowly than others. As a crystal grows, the slow-moving facet surfaces eventually define its shape.

How fast a given surface collects material and advances depends on the molecular structure of the crystal. If you could cut a crystal at a random angle and look at the individual molecules on the cut surface, you would find lots of dangling chemical bonds. Those surface molecules miss their former neighbors and are anxious to find new ones. Therefore, molecules that hit the surface are rapidly incorporated into it. Put another way, a randomly cut crystal surface is rough on the molecular scale, and rough surfaces accrue material quickly.

If you carefully cut your crystal along a facet plane, however, the surface would be relatively smooth on the molecular scale. The crystal structure is such that the facet surfaces have fewer dangling chemical bonds. In a sense, the molecules are arranged in straight rows, and if you cut along a row the cut will be cleaner. With fewer dangling bonds, free molecules are incorporated into the crystal at a slower pace.

If you start out with a small lump of a growing crystalline material, the molecular rough spots on the surface will incorporate new molecules quickly, so these surfaces will advance outward quickly. Meanwhile, the adjacent smooth surfaces will not advance so rapidly, and these slowly moving surfaces will broaden to form facets. Before long, only the slow-moving faceted surfaces are left, defining the shape of the growing crystal.

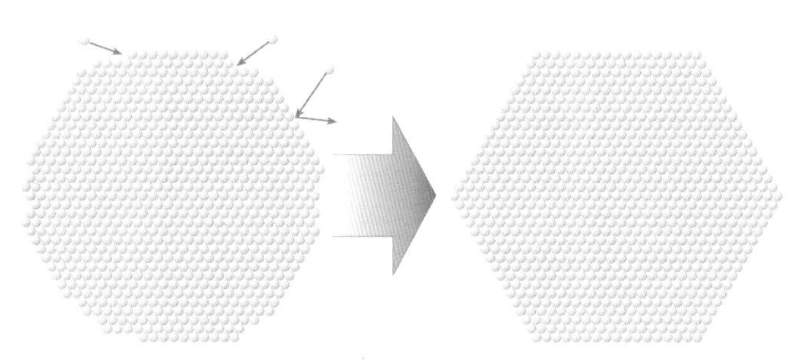

WHERE DO FACETS COME FROM?

Facets form as molecules are added to a crystal. If we start with a small, round crystal, molecules quickly attach themselves to the molecular steps on the surface, because that's where they will be most tightly bound. The smooth, flat, facet surfaces only accumulate molecules slowly, because molecules don't stick so readily there. After the fast-growing regions fill in, all that remains are the slow-growing facet surfaces, which then define the shape of the crystal.

THE MANY FACETS OF SNOWFLAKES
Faceting plays a major role in the growth of snow crystals. This snow crystal features several different facets.

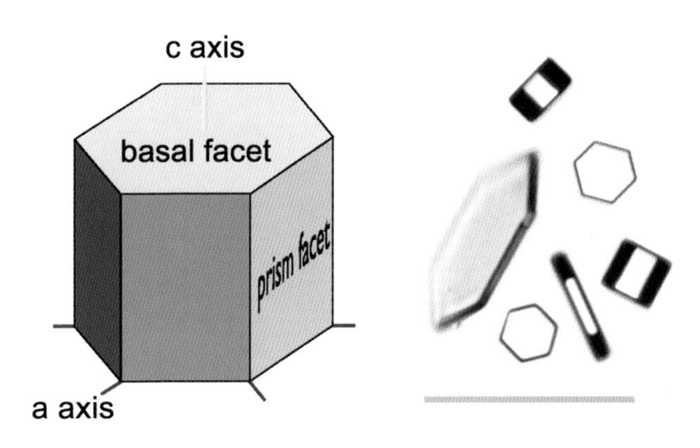

HEXAGONAL PRISMS
The most basic snow-crystal shape is the *hexagonal prism*, left, which includes two basal facets and six prism facets. The small, laboratory-grown snow crystals, right, are all shaped like hexagonal prisms. Depending on which facets grow faster, a hexagonal prism can become a thin plate or long column. The crystals shown are tiny, smaller than the width of a human hair. (Photo © Kenneth Libbrecht)

The molecular forces act locally and on a small scale, but long-range order and structure result. This is how the geometry of a molecule governs the geometry of a large crystal.

Hexagonal crystals like ice have two principal types of facet surfaces, the *basal* and *prism* facets. A perfectly faceted ice crystal is called a *hexagonal prism*, which has two basal facets on the top and bottom, and six prism facets forming the sides of the prism. Small snow crystals can often be found with this form. They may be short and broad like thin hexagonal plates, or tall and slender like columns. Whether plate-like or columnar, simple snow crystals have the same basic structure. The only difference is the relative growth rates of the two principal facets.

Why are there these two, and only these two, facet surfaces in an ice crystal? Well, that results from the structure of water molecules and how they connect together in a crystal. Other crystalline materials, made of different types of atoms and molecules, connect differently and form different facets. Salt, for example, has a cubic crystal structure and grows into box-like rectangular prisms. Other minerals can grow into a variety of structures—tetragonal, rhombohedral, and other forms. As Kepler deduced, ice is simply another crystalline mineral, one of the more common minerals on Earth.

The symmetry you see in a snowflake descends directly from the most fundamental mathematical symmetries of nature. The snowflake's hexagonal patterning derives from the structure of the ice-crystal lattice. The lattice structure in turn derives from the geometry of water molecules and how they connect. This is determined by the quantum mechanics of how atoms interact to form chemical bonds. The chain of reasoning quickly brings us to the most elementary laws of physics.

Chapter 4

Hieroglyphs from the Sky

 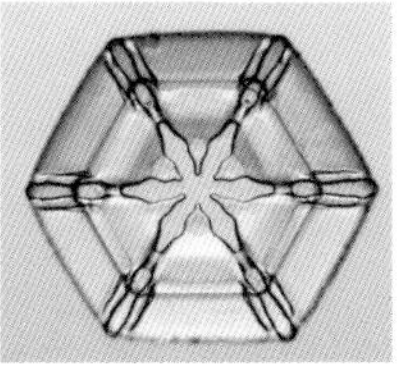 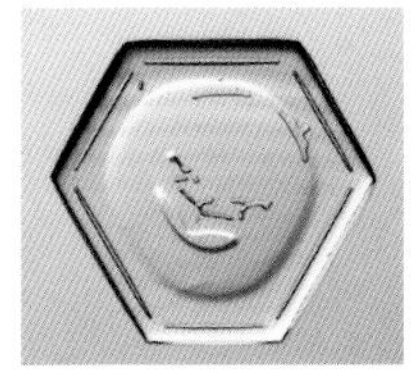

"A careful study of this internal structure not only reveals new and far greater elegance of form than the simple outlines exhibit, but by means of these wonderfully delicate and exquisite figures much may be learned of the history of each crystal, and the changes through which it has passed in its journey through cloudland. Was ever life history written in more dainty hieroglyphics!"
—*W. A. Bentley and G. H. Perkins, "A Study of Snow Crystals,"* Appleton's Popular Scientific Monthly, *May 1898*

The real puzzle of snowflakes is not their symmetry alone. The real puzzle is their combination of symmetry and complexity—the fact that snow crystals grow into such complex shapes that are *also* symmetrical. Just look at an elaborate snow star and it begs the questions: How do the six arms each develop the same ornate shape? How do the branches coordinate the intricacies of their growth?

This snowflake puzzle was largely worked out in the 1930s when Japanese physicist Ukichiro Nakaya made the first detailed and systematic study of snow crystals. Whereas Johannes Kepler pondered the science of snow crystals without so much as a magnifying glass, Nakaya carried out an

Symmetrical Complexity
Facing page: The true puzzle of snowflake formation lies in the snowflake's natural combination of symmetry and complexity.

active experimental investigation armed with the arsenal of new scientific tools available at the beginning of the twentieth century.

Nakaya received his university training in nuclear physics, but upon graduation he was unable to find a research position matching his training—a common predicament of many university graduates. He eventually secured a position at the University of Hokkaido in northern Japan, but the institution had no facilities for conducting nuclear research.

Inspired by Bentley's photographs, Nakaya turned his attention to the study of snow crystals, capitalizing on the abundant supply of raw material available during the long Hokkaido winters. In addition to observing and categorizing natural snow crystals, Nakaya gained tremendous insights by growing synthetic snow crystals in his laboratory. In this way, Nakaya was able to examine how snow crystals grew under controlled conditions, obtaining vital clues for understanding the mysteries of snow-crystal formation.

One of the first difficulties Nakaya faced in trying to create snow crystals was that the natural ones float freely in the atmosphere as they develop. And since they fall a long distance, they have time to grow to a substantial size. Nakaya could not duplicate this in his lab, however, because crystals falling from a short height did not grow large enough to be studied using available equipment.

To solve this problem, Nakaya sought to suspend individual growing crystals on fine string so he could watch each one develop. That way the growth time could be extended indefinitely. Unfortunately, the technique did not readily yield isolated snow crystals, but rather coatings of frost. Nakaya explored many different filaments in his quest to make isolated snow crystals. He tried various strings of silk, cotton, and other materials, as well as fine wires, and even a spider's web. All resulted in frost-like clusters of minute ice crystals, not at all like natural snowflakes.

Nakaya finally achieved success with, of all things, a rabbit hair. The natural oils on the hair discouraged the nucleation of ice and prevented the growth of large numbers of frost crystals. Instead, isolated snow crystals grew into forms that bore an excellent resemblance to those produced in nature. These were the world's first synthetic snow crystals.

GROWING SNOW

Japanese snow-crystal researcher Ukichiro Nakaya observes the formation of synthetic snow crystals in his refrigerated laboratory. (Harvard University Press)

SYNTHETIC SNOWFLAKES

One of the world's first synthetic snow crystals, left, growing on a rabbit hair in Ukichiro Nakaya's laboratory, as pictured in his 1954 book *Snow Crystals: Natural and Artificial*. A natural specimen that fell from the sky is at right. (Harvard University Press)

The Morphology Diagram

After discovering the unexpected value of rabbit hair, Nakaya grew many individual crystals at different temperatures and humidity levels. He observed how the morphology of each crystal—its detailed shape and structure—depended on the condition of the air in which it grew. From these observations of synthetic snowflakes, Nakaya made a major breakthrough in understanding the variety and symmetry of snowflake forms.

Nakaya found that a snow crystal's morphology was remarkably sensitive to its growth conditions, especially temperature. When he made snow crystals at a temperature of -2° Celsius (28° Fahrenheit), just a bit below freezing conditions, they grew into thin plate-like crystals. When the temperature was lowered to -5° C (23° F), long thin needles appeared. Continuing down to -15° C (5° F), thin plate-like crystals again grew. Below -25° C (-13° F), the morphology changed once more to a mixture of thick plate and columnar forms.

In addition to this unexpected temperature dependence, Nakaya also observed that the complexity of snow-crystal shapes increased with increasing humidity. At low humidities, crystals grew into simple hexagonal prisms, the most basic ice-crystal form. The prisms were generally more plate-like or columnar depending on temperature, but their morphologies were always simple.

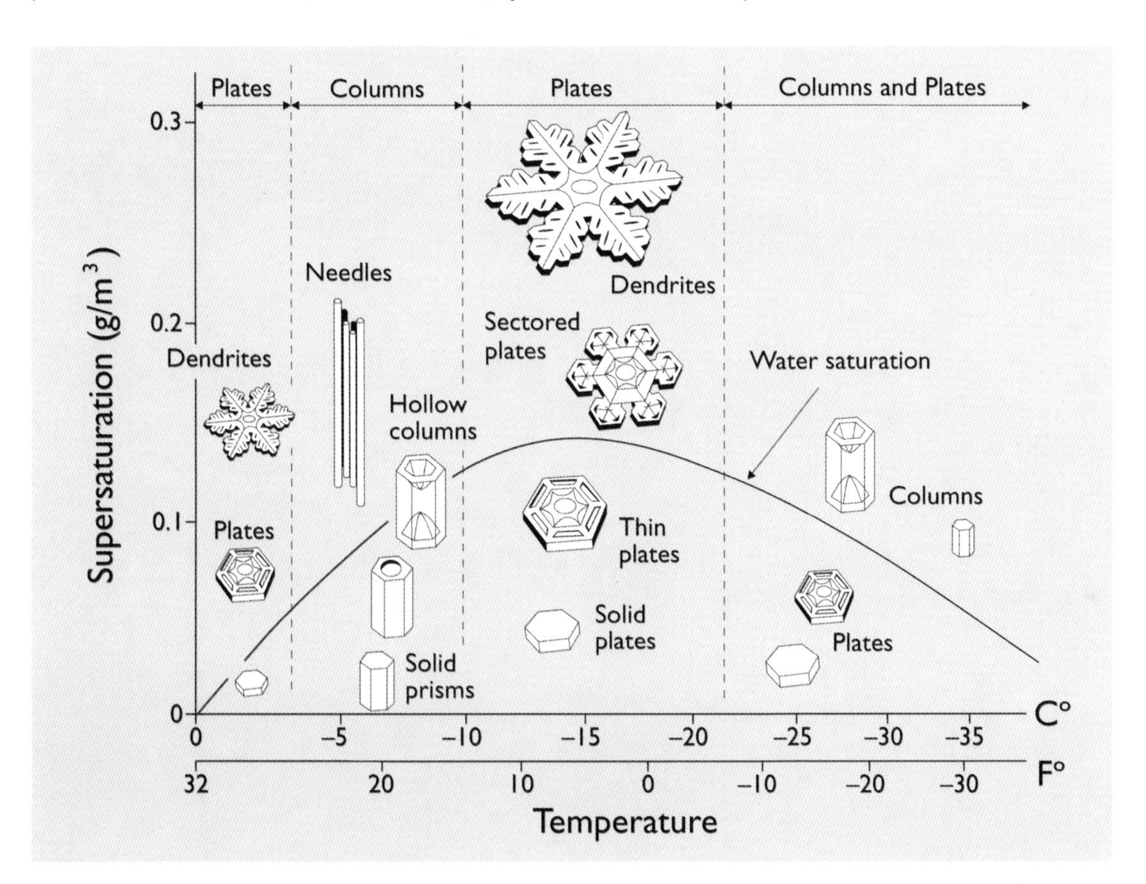

SNOW-CRYSTAL MORPHOLOGY

Ukichiro Nakaya's famous diagram shows the types of snow crystals that grow at different temperatures and humidity levels. Temperature mainly determines whether snow crystals grow into plates or columns, while higher humidities produce more complex structures. The *water saturation* line gives the humidity for air containing water droplets, as might be found within a dense cloud. This chart is adapted from a diagram by Y. Furakawa.

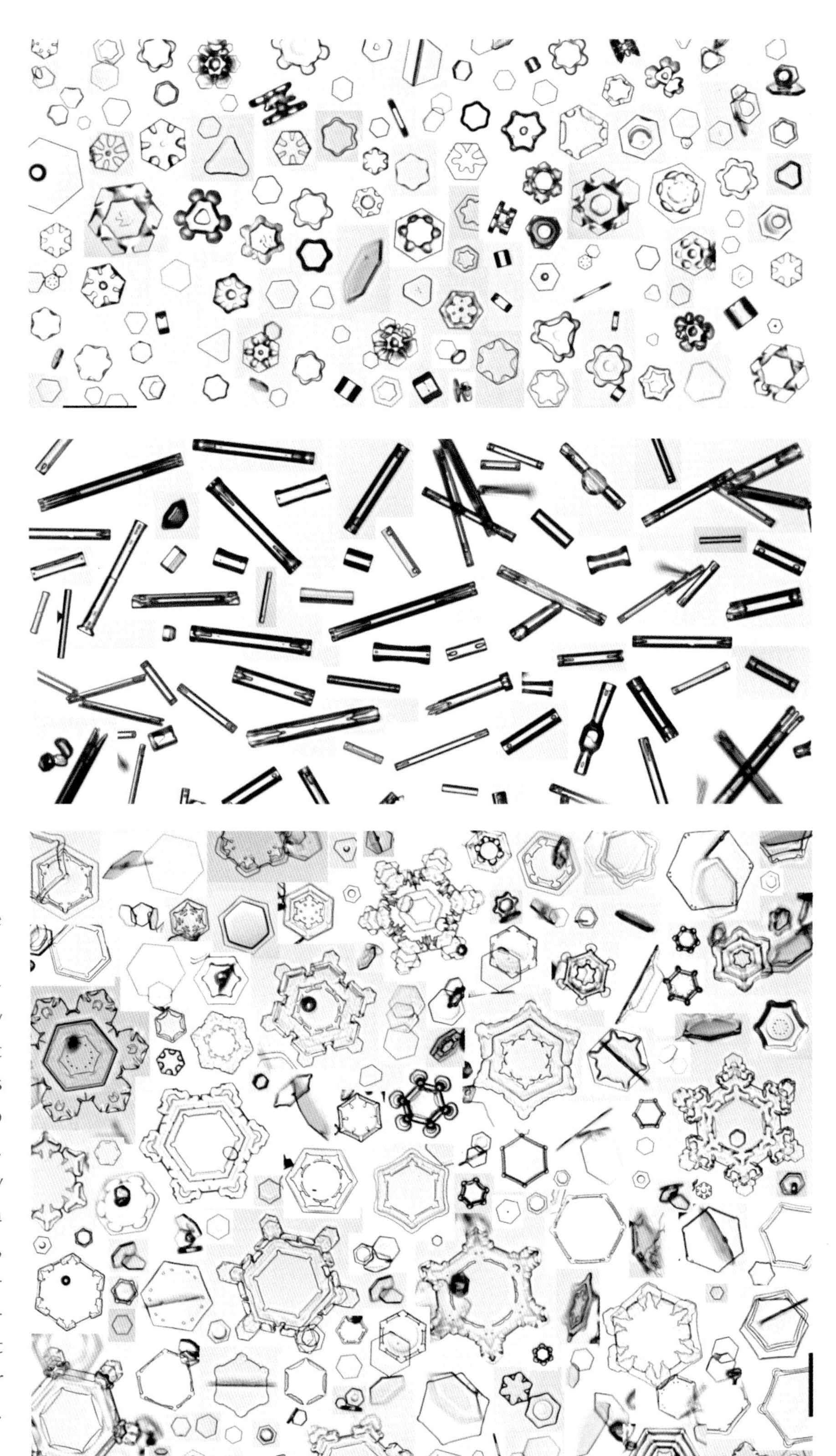

Homegrown Snow

These small crystals were grown inside a large cold-chamber in my laboratory. You can clearly see the dramatic differences in the crystal morphology with temperature from the plates that grew at -2° C (28° F), top, to columns at -5° C (23° F), center, and back to plates again at -15° C (5° F), bottom. The crystals grew as they fell freely within the chamber, landing after a minute or two on the chamber floor, where they were photographed. Because the chamber was small in comparison to a cloud, the crystals did not grow very large. Most are no larger than the thickness of a sheet of paper. (Photographs © Kenneth Libbrecht)

At moderate humidities, Nakaya observed the temperature variations becoming more accentuated. Plate crystals grew into larger and thinner varieties, while columnar crystals grew longer and more slender. Patterned plates appeared, as did hollow columns.

When the humidity was increased still more, to levels found inside dense mid-winter clouds, the most complex crystals emerged. Columns developed into sheath-like forms and clusters of thin ice-needles. Plate-like crystals blossomed into spectacular snow-crystal flowers, ornamented with intricate branched structures.

Nakaya displayed his measurements in a *snow-crystal morphology diagram*. This useful diagram is like a Rosetta Stone of snow-crystal growth, allowing one to interpret the form of a particular snowflake. The overall crystal shape, be it plate-like or columnar, reveals the temperature at which the crystal grew. The complexity of the structure indicates humidity. Snowflakes are then like hieroglyphs from the sky, carrying information about the conditions of the clouds in which they formed.

Complexity and Symmetry

From his studies of synthetic snow crystals, Nakaya was able to explain why snow crystals grow into such complex yet symmetric shapes, and how the six arms of a snow star coordinate their growth. The key to unlocking this mystery was the observation that snow-crystal growth is exceedingly sensitive to temperature and humidity.

Consider the life story of an individual snowflake—a large symmetrical snow star that you might catch on your mitten during a quiet snowfall. In the beginning, your crystal was born as a tiny nucleus of ice, and by its good fortune this nascent snowflake quickly grew into a well-formed single crystal of ice, a minute hexagonal prism.

While in its youth, fortune again smiled by placing the crystal in a region of the cloud where the humidity was just right and the temperature was a perfect -15° C (5° F). There the tiny crystal grew into a thin, flat hexagonal plate. In this early phase in its growth, the crystal shape was being determined mainly by faceting.

As it reached snow crystal adolescence, the crystal blew suddenly into a region of the cloud with high humidity. The increased water supply made the crystal grow faster, which in turn caused the corners of the plate to sprout small arms. Because the humidity increased suddenly, each of the six corners sprouted an arm at the same time. The arms sprouted independently of one another, yet their growth was coordinated because of the motion of the crystal through the cloud.

The crystal subsequently blew to and fro in the cloud while it grew, following the will of the wind. As it traveled, the crystal was exposed to different conditions. Since a snow crystal's growth depends strongly on its local environment, each change of the wind caused a change in the way the crystal grew. Again, each change was felt by all six arms at the same time, so the arms grew synchronously while the crystal danced through the clouds.

As the crystal grew larger and ever more ornate, it eventually became so heavy that it floated gently downward, out of the clouds to land on your mitten. The exact shape of each of the six arms reflects the history of the crystal's growth. The arms are nearly identical because they share the same history.

The precise morphology of each falling crystal is determined by its random and erratic motions through the atmosphere. A complex path yields a complex snowflake. And since no two crystals follow exactly the same path to the ground, no two crystals will be identical in appearance.

So where is the creative genius, capable of designing snow crystals in an endless variety of beautiful patterns? It lives in the ever-changing wind.

TWINS?

Identical twins? No. Close siblings? Definitely. These two snow crystals fell within a few minutes of one another, and clearly they both followed similar paths—plate-like growth in the beginning, followed by a period of branching growth.

Chapter 5

Morphogenesis on Ice

 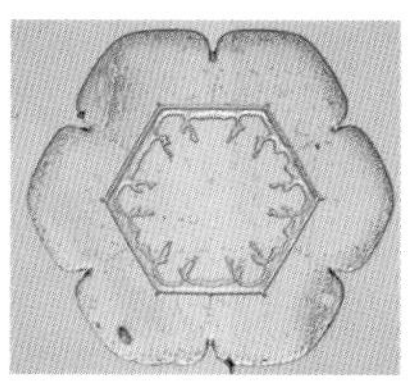 

"Everything is complicated; if that were not so, life and poetry and everything else would be a bore."
—*Wallace Stevens,* Letters of Wallace Stevens, *1967*

Complex objects—things with elaborate patterns or designs—are a familiar sight in our world. When I look around my living room I see a cluttered cacophony of complex stuff—books, lamps, chairs, clocks, and even a computer sitting in my lap. The source of all this complexity is obvious: it was manufactured, ultimately by human brains and hands. Even if I don't know how my computer works in detail, there's no real mystery surrounding its origin.

Looking outside my window at the trees and birds, I see a different kind of complexity, the complexity of life. This complexity is full of mystery. Just consider how a single small seed manages to produce a flower. What actually happens to create the structure and patterns in the leaves and petals?

Morphogenesis

Facing page: Snow-crystal morphogenesis occurs spontaneously—just add water.

A flower is an example of biological *morphogenesis*, the spontaneous creation of form—nature using chemistry and self-assembly to generate complexity. The genome guides the fabrication of proteins, which in turn guide the manufacture of other molecules. Some of the molecules coil themselves into various nanoscopic structures, which in turn self-assemble into larger structures and eventually into a flower. The mechanics of all this is almost unbelievably complicated. But it all happens spontaneously—just add water.

In his book *The Blind Watchmaker*, Richard Dawkins states, "The physicist's problem is the problem of ultimate origins and ultimate natural laws. The biologist's problem is the problem of complexity." While that's true enough, we physicists also dabble in complexity; we just pick easier problems. The trouble with trying to understand biological complexity is that there are no simple examples. All the really simple life forms—primordial sacks of slime and ooze—were eaten into extinction eons ago.

Fortunately there are plenty of examples of physical morphogenesis in systems that are far simpler than anything biological. Waves on the oceans, ripples on snowdrifts and sand dunes—all are relatively simple pattern forming systems in which complexity arises spontaneously.

My favorite example, of course, is the snowflake—the poster child of morphogenesis. Snowflakes often have exceedingly complex shapes, and the patterning appears spontaneously as snow crystals grow.

Faceting explains how the structure of the ice lattice is imparted onto a snow crystal's growth and form, so faceting explains a snow crystal's six-fold symmetry. But if the slow-growing facets were the whole story, then all snowflakes would look like simple hexagonal prisms. We need something more to explain why snow crystals fall to earth in such complex, lacy structures. There are endless variations of snow-crystal shapes, but each and every one is produced by the same simple process—water vapor condensing into ice. How does the simple act of freezing produce such elaborate structures?

The Branching Instability

Growth is the key ingredient for the generation of snow-crystal patterns. Left in isolation for a long time, an ice crystal will eventually turn into a plain hexagonal prism. When snowflakes are stuck in a snowbank, they tend toward this simple equilibrium structure, which is why stellar crystals quickly lose their shape once they hit the ground—the elaborate crystal designs are not stable. Ornate patterns appear only when a snow crystal is out of equilibrium, while it is growing.

A snow crystal grows by grabbing water molecules out of the air and incorporating them into itself. Water vapor molecules are assimilated into the existing ice lattice, which then increases in size. As long as the humidity is sufficiently high, the crystal will grow; there will be a flow of water from air to ice.

As a crystal grows, however, it consumes the excess water vapor around it, depleting the nearby air and reducing its humidity. To keep growing, water molecules from farther away must diffuse through the air into the depleted region near the crystal. This process takes time, so diffusion impedes the crystal's growth. Under such circumstances we say the growth is *diffusion limited*. The crystal development is governed by how quickly molecules can make their way to the crystal. Diffusion-limited growth often leads to branching.

Consider a simple hexagonal plate crystal as it floats through a cloud. Because the hexagon's six points stick out a tiny bit, water molecules are a bit more likely to diffuse to the points than to anywhere else on the crystal. The points then tend to grow a bit faster, and before long they stick out farther than they did before. Thus the points grow faster still. The growth becomes an unstable cycle: the points stick out a bit, they grow faster, they stick out more, they grow faster still.

This kind of positive feedback produces what is called a *branching instability*—even the tiniest protruding points will grow faster than their surroundings and thus protrude even more. Small corners grow into branches; random bumps on the branches grow into sidebranches. Complexity is born.

Instabilities like this are the heart of pattern formation, and nature is one unstable system heaped on top of another. The sun heats the air near the ground and the warm air rises—a connective instability that drives the wind, clouds, and all of our weather. The resulting wind blows on the surface of the ocean, making the ocean surface unstable, and waves are generated. The waves travel across the ocean, and when they run into a shallow beach the waves become unstable and break. Instabilities are responsible for many of the patterns you see in nature, including snowflakes.

Complex Patterns

Snow crystals form complex structures through a combination of faceting and branching. Faceting produces thin plates and flat edges while branching promotes the growth of more elaborate patterns.

Dendrites

The formation of snow-crystal *dendrites* is a good example of diffusion-limited growth and the branching that results. The word dendrite derives from the Greek *dendron*, meaning "tree" or "tree-like," referring to the complex structures these crystals develop. We see dendritic structures in snow crystals grown at nearly any temperature, as long as the humidity is high. But the most common snow-crystal dendrites are the fern-like crystals that grow at -15° C (5° F).

A dendrite occurs when the branching instability is applied over and over to a growing crystal. As the tip of a branch grows longer, random bumps on the sides grow into sidebranches. The sidebranches are often irregularly spaced, reflecting the random nature of the instability. If the dendrite is large enough, it will begin to show a self-similar, or fractal, structure, in that the branches have sidebranches, and these in turn have their own sidebranches, and so on.

If we want to change the character of the branching, we need only change the rate of diffusion. We can do this in the lab by changing the pressure of the air in which snow crystals grow. At lower pressures, water molecules diffuse more quickly, so growth is less limited by diffusion. Snow crystals then show less branching and are more faceted. When the pressure is low, they invariably grow as simple hexagonal prisms. On the other hand, at air pressures above one atmosphere, snow crystals show even more extensive branching than what nature normally provides. Alien planets, with atmospheres different from ours, may well exhibit different kinds of snowflakes.

DENDRITES

Above: These two snow-crystal dendrites were grown in my lab, the upper one at -15° C (5° F), the lower at -5° C (23° F). They were grown at high humidity, which resulted in extensive sidebranching. Each is just over 0.1 mm (0.04 inches) long. (Photographs © Kenneth Libbrecht)

POND CRYSTAL

Left: This exceptionally large, dendritic ice crystal grew on the surface of a still pond. Although it looks a bit like a snow crystal, it was formed by the freezing of liquid water, not water vapor. The angles between the various branches and sidebranches reveal that it is a single crystal of ice, just like a snow crystal. The dendritic structure results because the ice growth is limited by heat diffusion in the water. (Photograph © Bathsheba Grossman)

Designer Snowflakes

One of my favorite tricks for producing laboratory-grown snow crystals is to grow them on the ends of thin ice-needles. The basic idea was first discovered in 1963 in the laboratory of Basil Mason at Imperial College in London. Over the past few years, we've further improved the technique in my lab at Caltech, and we've calculated how the process works in detail.

We start with tiny frost crystals on the end of a wire in the middle of a chamber filled with moist air, and then apply 2,000 volts to the wire. In a few minutes, the small crystals almost spring to life as thin ice-needles begin growing rapidly outward. The voltage produces strong electric fields on the ice crystals and the fields attract water molecules in the air, thereby greatly accelerating the crystal growth. By adding some chemical vapors to the air, we can coax the needles to grow straight and true.

After the electric ice-needles have grown for a while, we remove the high voltage and the crystal growth goes back to normal. If the needles are at a temperature of -15°C (5°F), then plate-like snow crystals develop on the needle tips. With a bit of care, we can grow snow crystals that are quite similar to the natural variety and with many morphologies. By changing the temperature and humidity as a crystal grows, we can even make designer snowflakes with patterns of our own choosing.

ELECTRIC ICE-NEEDLES

A pair of thin needles of ice grew when the ice crystal was charged up to 2,000 volts. After removing the voltage, normal ice stars grew on the ends of the needles, right. (Photographs © Kenneth Libbrecht)

SNOWFLAKE QUARTET

Above: These four synthetic snow crystals were grown in my lab on the ends of thin ice-needles. (Photograph © Kenneth Libbrecht)

DESIGNER SNOWFLAKE

Left: This designer snowflake's pattern was determined by controlling the growth conditions. It was grown on the end of an ice needle that is out of sight directly behind the crystal. Tip to tip, the crystal is about 1.5 mm (0.06 inches). (Photograph © Kenneth Libbrecht)

The Morphological Balance

If you look at enough snowflakes, you can see that their growth is governed by a delicate balance between faceting and branching. If branching alone dominated the growth behavior, then snow crystals would all be dendritic, and they wouldn't be flat. They might look a bit like miniature ice tumbleweeds—round, with an open, ramified structure inside. If faceting alone dominated, snow crystals would all look like simple hexagonal prisms.

In most cases, neither faceting nor branching is completely dominant. It's the combination of both that gives a snow crystal its character. In a large stellar crystal, for example, the slow-growing basal facets give the crystal its overall flatness. The branching instability produces the crystal's complex fern-like structure, but the 60-degree angles between the branches are set by faceting. Both faceting and branching play important roles in these crystals.

You can get a feeling for how the balancing act between branching and faceting works by again considering the growth of a simple hexagonal prism crystal. When the crystal is small, diffusion is not an important factor. Water molecules readily diffuse the short distance from one end of a tiny crystal to the other, so the supply of water is essentially the same over the entire surface. In this case, the growth is not diffusion limited at all, so faceting determines the crystal shape. Extremely small crystals often look like simple hexagonal prisms for this reason.

As the prism becomes larger, diffusion starts to limit the growth, so the corners start to grow a bit faster than the centers of the crystal faces. But as soon as that happens, the facets will no longer remain exactly flat. When the face centers start to lag behind, their surfaces become slightly curved, exposing some extra molecular bonds. Since surfaces with exposed bonds accrue material more quickly than flat faceted surfaces, the faces are able to keep up with the corners, even though their water supply is lower.

For a while the forces of branching and faceting are held in balance, and the ice surface maintains just the right curvature. If it gets a bit too flat, then branching begins to kick in, causing the corners to grow faster, increasing the curvature. If the curvature is too great, the faces grow faster and catch up. A dynamic equilibrium is maintained automatically, and for a while the crystal keeps its simple faceted appearance. The facets are not precisely flat on the molecular scale, but they look flat because the curvature is so slight.

As the crystal grows still larger, however, the branching instability becomes an ever greater force. The faces become ever more curved and thus rougher on the molecular scale. Eventually the face centers become completely rough, and their growth is then limited only by diffusion. Soon after this happens, the faces will no longer be able to keep up the pace, and the hexagonal prism will sprout arms. Branching has won; instability kicks in.

The bottom line is that both faceting and branching are simultaneously important for determining snow-crystal structure. Furthermore, the interplay between these two growth mechanisms is complicated. It depends on temperature, humidity, and even the size and shape of the growing crystal. The delicate balance between these two forces gives snow crystals their tremendous diversity.

THE BIRTH OF BRANCHES
This series of photographs shows the growth of a simple plate-like snow crystal, demonstrating the transition from faceting to branching. Faceting dominates when the plate is small, yielding a hexagonal plate. When it grows larger, the corners of the hexagon stick out so far that branches form. (Photographs © Kenneth Libbrecht)

Remaining Puzzles

The interplay between branching and faceting makes it all but impossible to predict what any individual snow crystal will look like. But really this has to be the case; if the problem were simple, then snow crystals would not grow into such a stunning variety of shapes. The variety arises exactly because the growth is so sensitive to the ever-changing conditions inside a cloud.

The physics behind snow-crystal symmetry and faceting is well understood, and we have a pretty good handle on branching as well. These phenomena can be fairly complex when they play out, but the underlying mechanisms make sense. Much of what we see in snowflakes can be explained by these effects.

The morphology diagram, however, remains an unsolved puzzle. Nakaya first described this key to snowflake formation more than fifty years ago, and we still don't understand why snow crystals grow according to these rules. Why do plates grow so thin and large at -15° C (5° F)? Why do long slender columns grow at -5° C (23° F)? What mechanism causes the growth to change so dramatically over such a small temperature interval?

The eventual answers to these questions lie in the surface structure of ice. We know how water molecules are arranged *inside* an ice crystal, a process that has been measured and calculated for decades. But the molecules at the ice surface are not bound together as tightly as those in the interior. With different binding, the molecular structure of the surface is different from the interior. We currently don't understand the surface structure of ice in detail, or how it affects crystal growth.

So here we sit at the beginning of the twenty-first century and we cannot yet explain exactly why snowflakes are what they are. Snowflakes are full of surprises, and there are still some fundamental aspects of snowflakes we do not understand. A bit of mystery remains in these delicate ice structures.

GROWING SNOWFLAKE

This series of photographs show a growing snow star over a period of about fifteen minutes. (Photograph © Kenneth Libbrecht)

Chapter 6

SNOWFLAKE WEATHER

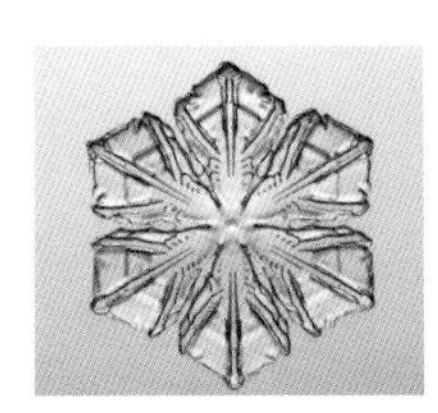

"When the temperature of the air is within a degree or two of the freezing point, and much snow falls, it frequently consists of large irregular flakes. . . . But in severe frosts, . . . the most regular and beautiful forms are always seen floating in the air, and sparkling in the sun-beams; and the snow which falls in general is of the most elegant texture and appearance."
—*William Scoresby,* An Account of the Arctic Regions with a History and Description of the Northern Whale-Fishery, *1820*

Snowflakes are being manufactured in the atmosphere at an astounding rate—around a million billion crystals each second. Every ten minutes that's enough snow to make an unstoppable army of snowmen, one for every person in the world. Over the Earth's history, some ten times the mass of the planet has floated down to its surface in the form of tiny ice crystals.

The nurseries that produce these vast numbers of snowflakes are nothing more than the clouds filling the sky on a winter's day. We can learn a

WINTER CLOUDS
Facing page: Even when the temperature is below freezing, most clouds are made of tiny water droplets. Snow occurs when some of the droplets in a cloud freeze and grow into snowflakes. (Photograph © Richard Hamilton Smith)

thing or two about snowflakes by taking a look at the clouds in which they are created: where they come from, what they are made of, and how they manage to make snow. Instead of staring up at the gray winter sky and asking whether it will snow today, ask why it snows at all.

Snow Clouds

If you wanted to create your own personal snowstorm, your first step would be to get some water and put it into the air as water vapor. For a good-sized cloud bank, you need a million tons or so of water, so plan on using more than a garden hose. In nature this is accomplished mainly through evaporation from oceans, lakes, and rivers.

Once the water vapor is in the air, the next step is to cool the air down. When a parcel of air is cooled, its relative humidity increases. With sufficient cooling, the relative humidity rises to 100 percent, at which point the air is *saturated* with water vapor. Cool the air further and the humidity rises above 100 percent, so the air becomes *supersaturated*. Supersaturated air is holding too much water, and before long the water vapor condenses to form a cloud of countless minute water droplets.

Each water droplet in a cloud requires a nucleus of some kind to start the condensation process. The nuclei are provided by microscopic dust particles. There is enough dust around even in clean air that clouds condense pretty quickly once the air is supersaturated. The cloud droplets can remain suspended indefinitely, as long as they are sufficiently small, simply because they fall so slowly.

When you look at a cloud, you are seeing the collective appearance of an enormous number of tiny water droplets. Clouds can be seen because cloud droplets scatter light. Water vapor itself is invisible, but water droplets are not.

Even if the air temperature is below water's freezing point of 0° C (32° F), wintertime clouds are usually made of *liquid* water droplets. The tiny cloud droplets do not freeze readily, and they must be brought several degrees below the normal freezing point before they solidify. Typically, droplets must be chilled to between -6° C (21° F) and -15° C (5° F) before they turn into ice. Droplets of very pure water can be chilled to nearly -40° C (-40° F) before they freeze. Water that has been cooled below the normal freezing point is said to be *supercooled*, and most winter clouds are made of supercooled water droplets, which is essential for the formation of snowflakes.

Doing Your Part

There is a bit of you in every snowflake. That's because even you, right this moment, are making a contribution to the atmospheric water supply. Water is evaporating from your skin, plus you are putting water right into the air every time you exhale. In fact, you personally put so much water into the air that some of your water molecules almost certainly made it into the snowflakes pictured in this book.

You exhale roughly a liter of water per day into the atmosphere, and most of this water rains or snows back down again within about a week's time. The total global precipitation is about 1,000,000,000,000,000 (one quadrillion) times greater than the amount of water you exhale, so your impact on the weather is pretty minor.

But even if you contribute only one quadrillionth of the total water content in a snowflake, that is still about 1,000 molecules. It depends on how well things are mixed in the atmosphere, but there are probably, very roughly, about a thousand of your molecules captured in every snowflake picture. Thank you for your contribution—and keep up the good work.

Lake-Effect Snow

Large freshwater lakes are excellent sources of water vapor for snow clouds. Water evaporates readily from unfrozen lakes in wintertime, and a good wind can lift a huge amount of water from the surface of a lake. Often this water very quickly gets dumped as snow, and some of the heaviest snowfalls in the world are recorded just downwind of large lakes. This phenomenon is called *lake-effect snow*. (Photograph © Richard Hamilton Smith)

Ice Nucleation

Water droplets in clouds do not freeze readily because in a sense they don't know how. The molecules in a liquid droplet are disordered, jostling against one another in a chaotic jumble—the perpetual dance that comes from thermal motion. Water molecules need to become ordered into a crystalline lattice to freeze, but they have a hard time settling down on their own.

Water usually freezes only when there is some nearby surface to guide the molecules into their ordered state. If there happens to be some ice nearby, then the liquid molecules just above the ice surface have a molecular template to follow. So when ice is already present, water will freeze at the norm of 0° C (32° F).

In the absence of existing ice, water can be supercooled; something besides water is needed to jumpstart the freezing process. In a snow cloud, dust actually does double duty. First, a dust grain nucleates the formation of a liquid droplet, and then the captive dust particle provides the necessary template to nucleate freezing.

All dust is not created equal. How well a dust particle promotes freezing depends on what kind of template it provides. A good ice nucleus is one with a molecular template that closely resembles ice. A supercooled cloud droplet may freeze around -10° C (14° F) if it contains an average dust particle, or perhaps as high as -6° C (21° F) for the best ice nuclei.

People are always looking for ways to best nature, and various types of artificial dust have been found that work better than natural dust at nucleating ice in cloud droplets. Silver iodide smoke can nucleate tiny ice crystals up to -4° C (25° F), two degrees higher than the best natural dust. Silver iodide has a lattice structure that differs only slightly from ice; thus it provides an excellent template to nucleate freezing.

Perhaps Kurt Vonnegut received some inspiration for his fictional ice-nine from all this; the nucleation properties of silver iodide were discovered in 1946 by his brother, meteorologist Bernard Vonnegut.

Ice Nucleation and Weather Modification

Mark Twain once complained, "Everyone talks about the weather, but nobody ever does anything about it." Weather modification is probably the ultimate response.

After it was discovered that silver iodide was more effective than natural dust as an ice-nucleating agent, a substantial effort was launched to accelerate the process of precipitation through *cloud seeding*. The idea stems from the fact that ice nucleation often triggers rainfall in temperate climates, even in the summer.

The trick to getting rain out of a cloud is to make some of the cloud droplets grow large enough to fall; otherwise the water simply stays up in the cloud. If the cloud tops are below freezing, then once ice particles appear they will grow into snowflakes. If they grow large enough, they will fall and they won't stop falling if they melt on the way down. Falling droplets also collide with cloud droplets as they fall, further increasing their size.

This scenario is a common occurrence in clouds. Rain is often made from snowflakes that were produced high in the clouds and melted as they fell. In such cases, ice nucleation is the first step in producing rain.

When silver iodide smoke is dispersed inside a cloud, smoke particles attach to cloud droplets and make them freeze more quickly. It's a small effect, like lowering the cloud temperature a few degrees, but it can sometimes tip the balance and coax a cloud to rain.

Cloud seeding mostly fell out of favor during the 1960s. It can only provide a small perturbation to the natural processes in clouds and some cloud-seeding practitioners promised too much. There were also some interesting legal tangles that resulted when rain fell where it wasn't supposed to. Cloud seeding is still being researched, and may someday become a useful method of weather modification.

Starting a Snowfall

Ice nucleation is a key step in snow clouds, since this is how water droplets freeze to become tiny, nascent snowflakes. Not all droplets freeze at once, since the temperature inside a cloud varies from place to place and the freezing point of each droplet is determined by the dust it contains. The nucleation process may take hours, days, or even weeks as a cloud drifts through the sky.

Once an individual droplet freezes into a microscopic ice particle, it grows by condensing water vapor from the air, thus forming a full-fledged snow crystal. The snow crystal grows as it floats through the cloud. After some tens of minutes the snow crystal will have grown large enough that gravity pulls it to the ground.

The additional water vapor needed to grow a snow crystal is provided indirectly by the remaining liquid cloud droplets. These will slowly evaporate away, putting water vapor into the air, which is then consumed by the growing snow crystals. During a snowfall there is a net flow of water—from droplets to air, and from air to ice crystals. That is essentially how a cloud freezes, turning its liquid water droplets into solid snowflakes.

If the cloud droplets freeze gradually, then the cloud

N1a Elementary needle	C1f Hollow column	P2b Stellar crystal with sectorlike ends
N1b Bundle of elementary needles	C1g Solid thick plate	P2c Dendritic crystal with plates of ends
N1c Elementary sheath	C1h Thick plate of skeleton form	P2d Dendritic crystal with sectorlike ends
N1d Bundle of elementary sheaths	C1i Scroll	P2e Plate with simple extensions
N1e Long solid needle	C2a Combination of bullets	P2f Plate with sectorlike extensions
N2a Combination of needles	C2b Combination of columns	P2g Plate with dendritic extensions
N2b Combination of sheaths	P1a Hexagon plate	P3a Two branched crystal
N2c Combination of long solid colmns	P1b Crystal with sectorlike branches	P3b Three-branched crystal
C1a Pyramid	P1c Crystal with broad branches	P3c Four-branched crystal
C1b Cup	P1d Stellar crystal	P4a Broad branch crystal with 12 branches
C1c Solid bullet	P1e Ordinary dendritic crystal	P4b Dendritic crystal with 12 branches
C1d Hollow bullet	P1f Fernlike crystal	P5 Malformed crystal
C1e Solid column	P2a Stellar crystal with plates at ends	P6a Plate with spatial plates

P6b Plate with spatial dendrites	CP3d Plate with scrolls at ends	R3c Groupel-like snow with nonrimed extensions
P6c Stellar crystal with spatial dendrites	S1 Side planes	R4a Hexagonal groupel
P6d Stellar crystal with spatial dendrites	S2 Scalelike side planes	R4b Lump groupel
P7a Radiating assemblage of plates	S3 Combination of side planes, bullets, and columns	R4c Conelike groupel
P7b Radiating assemblage of dendrites	R1a Rimed needle crystal	I1 Ice particle
CP1a Column with plates	R1b Rimed columnar crystal	I2 Rimed particle
CP1b Column with dendrites	R1c Rimed plate or sector	I3a Broken branch
CP1c Multiple capped column	R1d Rimed stellar crystal	I3b Rimed broken branch
CP2a Bullet with plates	R2a Densely rimed plate or sector	I4 Miscellaneous
CP2b Bullet with dendrites	R2b Densely rimed stellar crystal	G1 Minute column G2 Germ of skeletal form
CP3a Stellar crystal with needles	R2c Stellar crystal with rimed spatial branches	G3 Minute hexagonal plate
CP3b Stellar crystal with columns	R3a Groupel-like snow of hexagonal types	G4 Minute stellar crystal G5 Minute assemblage of plates
CP3c Stellar crystal with scrolls at ends	R3b Groupel-like snow of lump type	G6 Irregular germ

SNOW-CRYSTAL CLASSIFICATION SYSTEM

Many of the early snow-crystal observers, such as William Scoresby and Wilson Bentley, attempted to classify what they observed into types of snowflakes and snow crystals. These efforts were greatly improved by Ukichiro Nakaya, who produced a detailed classification based on his observations. Nakaya's table was subsequently extended by C. Magano and C. W. Lee with eighty classifications.

will produce a light snowfall. If the freezing of many droplets is triggered quickly—by the cloud temperature falling suddenly, for example—then the snowfall could be heavy.

The character of a snowstorm is governed by many factors—the cloud bank's size and density, temperature, how quickly the cloud droplets freeze, and so on. Predicting the precise behavior of the weather at your house is like predicting the exact shape of a snowflake—it is nearly impossible because so many factors are involved.

Homemade Weather

You don't have to look to the clouds to see weather-like phenomena. Many of the same things happen right in your backyard. For example, when the sun goes down, the ground cools, the air cools, and the relative humidity goes up. Above 100 percent humidity, water vapor condenses as dew droplets on your lawn. If the temperature is low enough, the droplets soon freeze into frost crystals, which can grow larger by condensing additional water vapor from the air. The process is basically the same as what happens in clouds, except the blades of grass provide nucleation sites.

You also witness condensation into droplets by simply breathing on a cold day. You provide warm moisture-laden air when you exhale, and your breath is rapidly cooled by the surrounding air. The water vapor condenses into a small cloud of water droplets, so you can see your breath. As your small cloud disperses, it mixes with the surrounding air, and the droplets evaporate again because the humidity of the air is less than 100 percent. So your breath disappears.

If you don't want to venture to your backyard, you can see these same weather phenomena right in your kitchen whenever you heat up your teapot. The water vapor coming out of the spout supersaturates the air, and condensing droplets make the steam cloud you see billowing from the teapot. Steam clouds, like atmospheric clouds, are made of water droplets.

If you place a glass plate in the supersaturated air produced by your teapot, water droplets will condense on the glass. If your kitchen were cold enough, frost crystals would appear. If your kitchen were *really* cold, you might even create tiny snow crystals in the air right above the teapot. For this to happen, however, the temperature would have to be in the neighborhood of -40° C (-40° F), a mighty cold kitchen even by North Dakota standards. In arctic climates, a neat trick is to toss some nearly boiling water from a pot into the outside air on a blisteringly cold night. *Whoosh*—instant snowflakes!

ARTIFICIAL SNOW
Since making snow from water vapor is a slow process, artificial snow is made by rapidly freezing liquid water-droplets. A common method is to shoot water and pressurized air out of a row of nozzles surrounding a large fan. The compressed air breaks the water stream into tiny water droplets and the expansion further cools the water droplets so they freeze. These are essentially sleet particles, not snowflakes, because they are made from freezing liquid water. Artificial snow does not match fresh powder on the ski slopes because sleet particles do not have the open dendritic structure of natural snowflakes. (Spirit Mountain Recreation Area)

Chapter 7

A Field Guide to Falling Snow

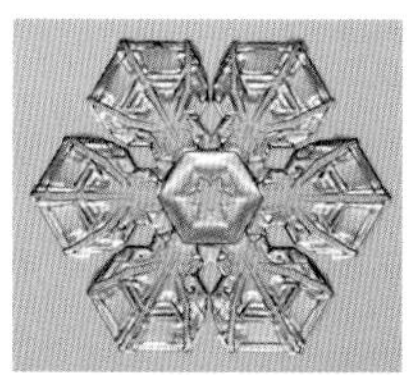
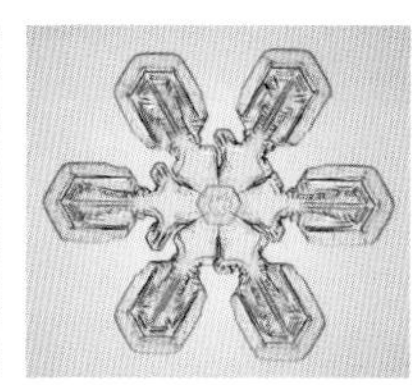

"The most beautiful thing we can experience is the mysterious. It is the source of all true art and science. He to whom this emotion is a stranger, who can no longer pause to wonder and stand rapt in awe, is as good as dead: his eyes are closed."
—*Albert Einstein,* What I Believe, *1930*

Snowflakes vary greatly in their patterns, yet not all snowflake patterns are possible in nature. There is an order to the diversity due to the rules governing snow-crystal growth. Some shapes appear frequently and can be found in most snowfalls. Others are uncommon or even rare. This chapter presents a field guide to falling snow, a tour of the ordinary and unusual members of the snowflake menagerie, annotated with an explanation of when and why the different types grow.

DIVERSITY AND ORDER
Facing page: The diversity of snow-crystal growth actually follows an order, and some crystal shapes appear frequently, others rarely.

Diamond Dust

We'll start our tour with the smallest snow crystals, known in the aggregate as *diamond dust*. Diamond-dust crystals are typically only a few tenths of a millimeter in size, a few times the thickness of a sheet of paper. Snowstorms often produce some of these smaller crystals, especially on colder days. The tiny crystals often go unnoticed because they're nearly invisible to the naked eye.

Diamond-dust crystals are most common high above ground, where the atmosphere is colder and drier than at lower altitudes. Cirrus clouds are wispy high-altitude clouds made of diamond-dust crystals. As with low-altitude clouds, the minute cirrus crystals are not large enough to fall, so they remain suspended in the atmosphere, following prevailing winds.

If the sky is laced with thin cirrus clouds on an otherwise sunny day, you can observe diamond-dust crystals indirectly by the way sunlight is reflected and refracted by the hexagonal ice prisms. Each crystal sparkles in the sunlight as it tumbles about, and the combined effect of millions of sparkles produces an *atmospheric halo*.

The simplest halo is one that occasionally encircles the sun with a 22-degree angle between the sun and any point on the halo. Although a suspended ice prism can deflect sunlight by any angle, a 22-degree deflection angle is especially likely. If you average the deflections of many randomly oriented crystals, a bright halo appears 22 degrees from the sun. The halo's intensity depends on the quality of the cirrus crystals; the most pronounced halos occur when the crystals are nearly perfect hexagonal prisms.

The 22-degree halo is fairly common in winter, and can be seen pretty much anywhere the climate is cold enough to require ownership of a snow shovel. When thin high clouds are present, halos appear around the sun by day and the full moon by night. Many people don't notice these halos simply because they don't look up.

Sun dogs, also known as *22-degree parhelia*, are related to the 22-degree halo. Sun dogs look like diffuse spots of light on either side of the sun, and they only appear when the sun is low in the sky. They are often colorful, although the colors are not as vivid as in a good

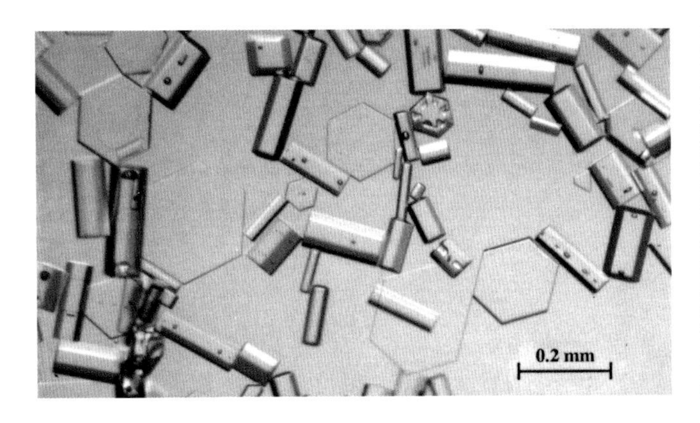

ANTARCTIC SNOWFLAKES
These diamond-dust snow crystals fell at the South Pole. They grew slowly in the dry Antarctic air, thus becoming nearly perfect hexagonal prisms. Most are only about 0.2 mm (0.008 inches) in size. Cirrus clouds are made of similar or even smaller ice crystals. (Photograph © Walter Tape, *Atmospheric Halos*)

ATMOSPHERIC HALOS
This dazzling halo display was recorded near the South Pole. The picture was taken using a wide-angle fish-eye camera lens, and a sign was used to block the sun so the rest of the display could be seen. The two brighter spots on either side of the sun are *sun dogs* and the circle passing through them is the *22-degree halo*. These two features can frequently be seen in less-remote locations; the numerous other arcs and halos are only rarely seen, even in the Antarctic. (Photograph © Walter Tape, *Atmospheric Halos*)

rainbow. Sun dogs appear when the suspended ice crystals are oriented with their basal faces parallel to the ground. This can happen when falling crystals are aligned by the aerodynamics of their motion, and good alignment is rare. Better-oriented crystals make better sun dogs.

There are many variants of atmospheric halo phenomena, with names like tangent arcs, Parry arcs, circumscribed halos, circumzenithal arcs, sun pillars, and others. Most of them are rare, since they require specific crystal arrangements—well-formed plate or columnar crystals, with different orientations in the sky.

The best place to find copious diamond-dust crystals, along with their atmospheric halos, is at the South Pole. Not only is it always bitterly cold at the Pole, the air is also dry—ideal conditions for producing small hexagonal prism crystals and spectacular halo displays.

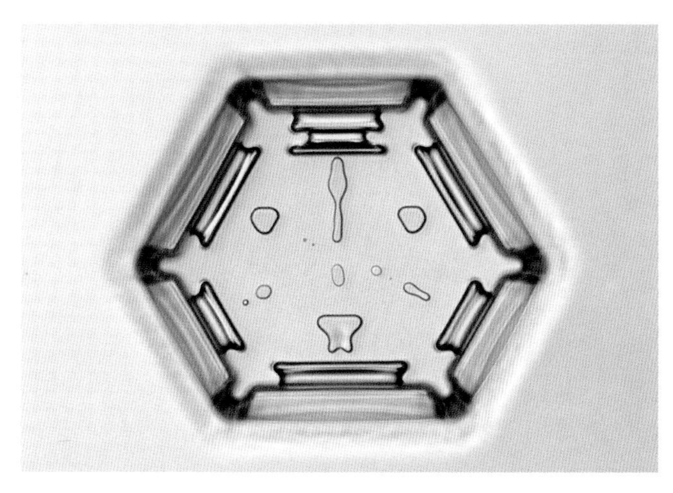

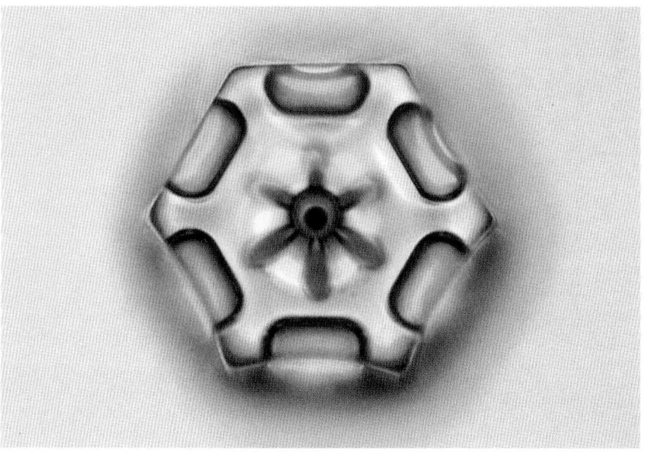

Stellar Dendrites

The largest snow crystals are the *stellar dendrites*—thin, flat crystals characterized by extensive sidebranching. In the most extreme cases they have a leafy, almost plant-like appearance.

The branches of a stellar dendrite grow rapidly in high humidity, and these crystals can reach sizes of 5 mm (0.20 inches) or more when conditions are ideal. The largest I've heard about in the wild was a specimen measuring 9 mm (0.36 inches) from tip to tip, recorded at Lone Butte, British Columbia. The larger crystals are fragile, so they break up in the slightest breeze. This puts an effective size limit on what you can find falling from the sky. In the lab, free from air currents, I once grew a specimen spanning 25 mm (1 inch) from tip to tip.

You may have heard reports of giant snowflakes as large as your fist, or perhaps even larger. These invariably are not single crystals, but loosely packed aggregates of crystals. When falling snow is wet and sticky, these puff-ball snowflakes can indeed grow quite large.

In spite of their complex shapes, stellar dendrites are still single crystals; the molecular ordering is the same throughout, from one tip to the other. The angles between the various branches are all multiples of 60 degrees, and you can see that the sidebranches are all parallel to their adjacent main branches. It may be hard to believe that the molecules at one end of such a complex crystal are still lined up with respect to the molecules at the other end, but the ridges and facets show that this is indeed the case.

Stellar dendrites typically exhibit six symmetrical main branches, each adorned with a collection of irregularly spaced sidebranches. The main branches are symmetrical because they sprout simultaneously from the six corners of an initial hexagonal ice prism. The sidebranches form more randomly, although sometimes symmetrical sidebranches are triggered by a sudden change in temperature or humidity.

Stellar dendrites form in high humidity when the air temperature is near either -2° C (28° F) or -15° C (5° F).

The higher of these temperatures is so close to ice's melting point, however, that large stellar dendrites usually don't form readily when conditions are so warm. It's possible, but if the temperature fluctuates a bit too high, the crystals melt; if the temperature fluctuates a bit too low, the growth briefly becomes more columnar and spoils the flat stellar shape.

The largest, most highly structured stellar dendrites always form when the cloud temperature is around -15° C (5° F). If the temperature is higher or lower, or if the humidity is not so high, the branching will be less pronounced.

Although complex branching is the most distinctive trait in stellar dendrites, these crystals are also remarkably thin and flat; the diameter-to-thickness ratio of a large crystal can be as high as 100:1. The flatness arises simply because the basal facets grow much more slowly than the prism facets at -15° C (5° F).

MONSTER CRYSTAL
This stellar dendrite is about as large as you're likely to find falling from the sky. At 7.2 mm (0.28 inches) from tip to tip, the crystal is roughly the size of a standard pencil eraser. Leafy specimens like this one are often called *fern* crystals for obvious reasons.

PAPER THIN
Stellar dendrites are remarkably thin, flat crystals. They are often less than 0.05 mm (0.002 inches) thick—half as thick as a sheet of paper.

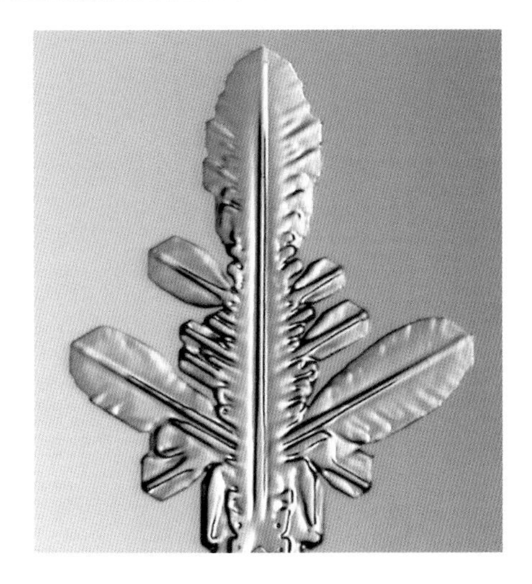

Grooves and Ridges

The branches of stellar dendrites often contain thin ridges running along their length, flanked by grooves. A ridge typically develops as a simple linear bump near the tip of a growing snow-crystal arm. As the ridge grows taller and the plate grows thicker, diffusion hinders the growth in the corner between the ridge and plate. Soon, the ridge is flanked by a symmetric pair of grooves in the ice.

Sectored Plates

When I go snow-crystal watching, I always hope I'll find some fine *sectored plates*, also called broad-branched crystals. These are thin, flat, plate crystals that arrive in an amazing diversity of shapes and sizes. Their identifying characteristic is the thin ridges that divide the crystals into sectors.

The simplest example is a plain hexagonal plate with ridges dividing it into six equal pieces, but this particular crystal type is exceedingly rare. Sectored plates more often appear as extensions on the ends of dendrite arms, as extensions of other plates, or in endless other combinations. It's the wonderful variety of these patterned plates that gives them such appeal.

Sectored plates are not especially rare, but they only form in conditions Goldilocks would appreciate—not too hot, not too cold, not too much or too little humidity. These crystals are the first cousins of the stellar dendrites: they are large, thin plates that only form right around the magic temperature of -15° C (5° F). Just one or two degrees away and the basal surfaces grow too rapidly, yielding thicker plates with a blockier appearance. It won't make much difference on your shovel, but the thinnest plates make the most attractive snow crystals.

Humidity defines the difference between a sectored plate and a stellar dendrite. If the humidity is high, dendritic branching occurs and the crystal becomes a stellar dendrite. If it's a bit lower, the branching will be less and sectored plates will appear. But if it's too low, the plates will not grow so fast and thin, again yielding blockier crystals.

Having all the conditions just right for making sectored-plate snow crystals is unusual, so on most days they are rare and the plates are not large. But when conditions are right, they're right. Sometimes, maybe just for a brief period, a snowfall can yield a large crop of excellent sectored-plate crystals. And because their growth is so sensitive to environmental conditions, a crop of sectored plates shows an exquisite diversity in appearance.

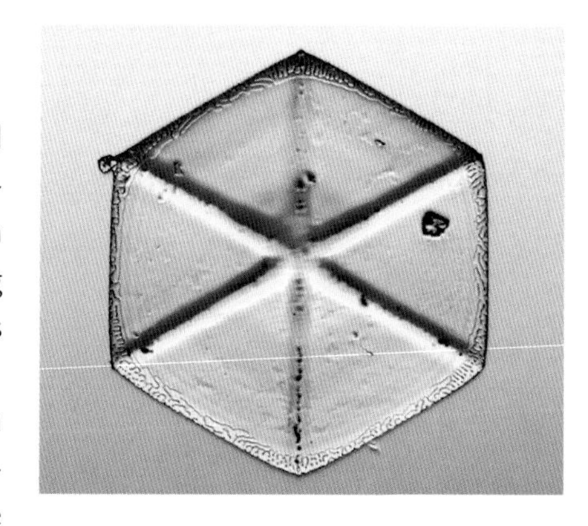

SIMPLE SECTORED PLATE

Hexagonal sectored plates are small and uncommon. This tiny example is only 0.3 mm (0.01 inches) in size.

If you have the inclination to slice up a snow crystal, with some care you can examine the cross section of the ridges. Near a growing tip, the ridge starts out as a simple linear bump, but farther from the tip the ridges are often flanked by small grooves. The grooves appear because the ridge grows taller and the plate grows thicker, but diffusion impedes the growth in the corner between ridge and plate.

A curious feature of sectored-plate crystals is that the ridges are sometimes curved or at odd angles, particularly with sectored plates growing on the ends of dendritic arms. It almost looks like the crystal structure has been bent, but this is not the case. The ice molecules are still all lined up with the usual crystalline symmetry.

A ridge is created at the corner between adjacent facets on the plate, and thus the ridge delineates the history of where the corner was at previous times. Adjacent facets don't always grow at the same rate, so the ridges can grow in some odd directions. If the relative growth rates of the facets changes with time, the ridges can become curved.

The ridges in a sectored plate in a sense depict its lifeline. The shapes of the ridges indicate how conditions changed as the crystal grew. It's all a matter of knowing how to decipher the snow-crystal hieroglyphics.

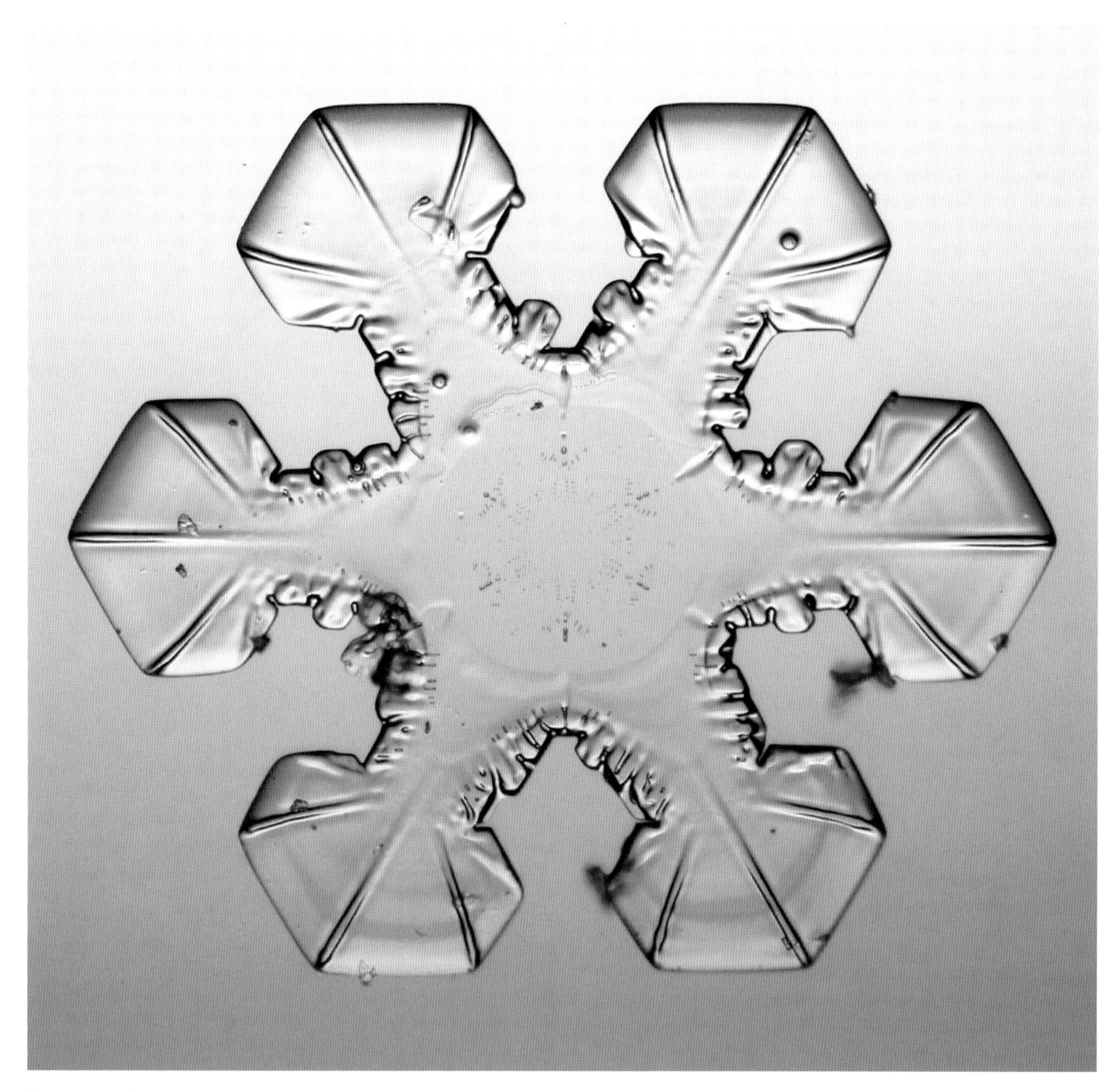

Ridged Plates

This branched crystal has sectored plate extensions with well-defined ridges. Similar ridges can often be seen in all types of stellar snow crystals. Many of these crystals have undergone some evaporation before being photographed, so the facets are not as sharp as they were when the crystals were growing.

Curved Ridges

When growth conditions change with time, sectored plates often develop curved ridges.

Columns and Needles

Columnar and needle forms are the forgotten members of the snow-crystal family. I daresay you will not find them depicted in holiday advertisements at your local shopping mall. In the real world, however, these crystals are common, and many snowfalls consist primarily of ice columns and needles.

Columnar crystals result simply when the basal surfaces advance faster than the prism surfaces. The simplest example is a long, slender hexagonal prism, shaped like a standard wooden pencil. Columns generally grow at temperatures around -5° C (23° F) and also at temperatures below a frigid -25° C (-13° F). Long columns look a bit like thin ice needles, and these grow only in a narrow temperature range around -5° C (23° F) when the humidity is high.

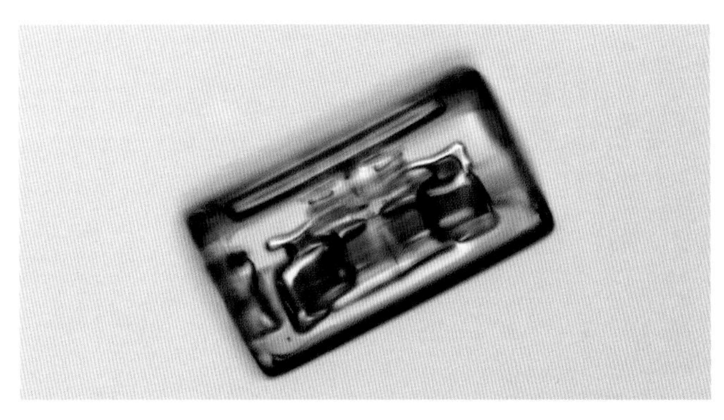

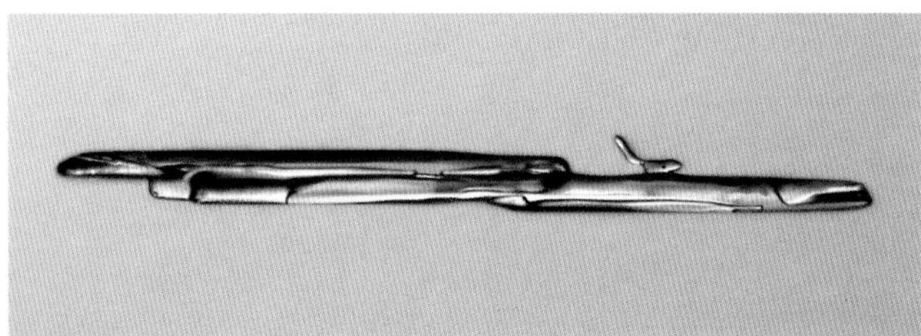

FROM BLOCKS TO NEEDLES

Above: Columnar snow crystals are common. They range from short, cylindrical blocks to long, thin needles. The small block crystal here is 1.8 mm (0.07 inches) long.

HOLLOW COLUMNS

Right: A close look at this frozen pillar reveals that it is partially hollow inside.

Hollow Columns

The growth of columnar crystals is controlled by the complex interplay between faceting and branching. When the crystals are small and the humidity is low, then faceting dominates. In this case, the crystals grow into simple hexagonal prisms.

When a hexagonal prism grows larger, diffusion becomes a factor and the branching instability takes on a new role. The prism facets continue to advance slowly, and they remain mostly faceted. But the fast-moving basal facets undergo changes. The basal edges grow more rapidly than the face centers, and soon a column will start to hollow. *Hollow column* crystals are common when the cloud temperatures are near -5° C (23° F).

The degree of hollowing depends on growth conditions and crystal size. Some crystals develop long, thin hollows and become almost sheath-like. Other crystals are stubbier, with shallow hollows. Sometimes a prism begins to hollow and then conditions change and the faces fill in. When this happens, it often leaves an enclosed hollow region inside the crystal—essentially a bubble within the ice.

Hollows and bubbles are not formed by removing ice from a crystal. Rather a hollow region forms when the edges of a crystal grow more rapidly than the face center; likewise, a bubble forms when a cap grows over a hollow. Snow crystals are sculpted by selectively adding material, rather than removing material.

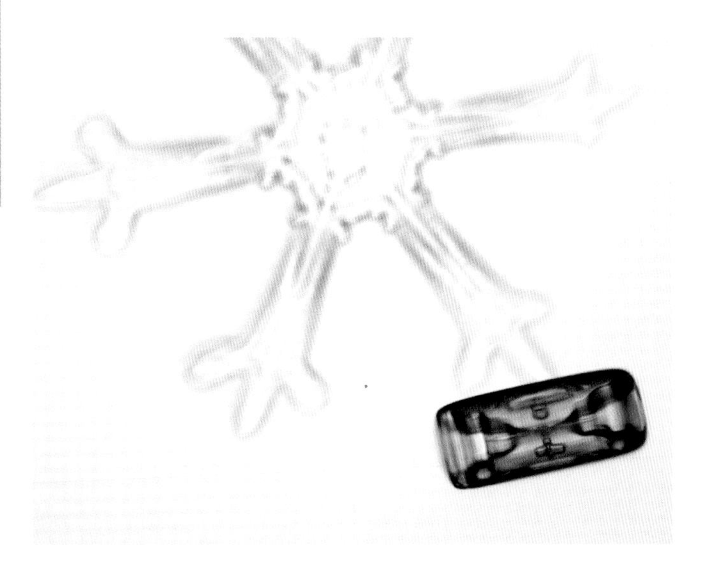

On the other hand, evaporation does often remove material from snow crystals after they've stopped growing. Sometimes this happens in flight, and it certainly occurs once they hit ground. The crystal edges often appear a bit rounded from evaporation. We see this frequently with columns because evaporation occurs more rapidly when the temperature is close to ice's melting point. At colder temperatures, the facets are more robust.

Needles and Bullets

If the humidity is high and the temperature is close to -5° C (23° F), then the rim of a hollow column can become so thin that its growth is also unstable. Some parts of the rim grow faster than others, and these parts soon grow out into a number of needle-like extensions.

This is yet another manifestation of the branching instability. Once one part of the rim grows ahead of the rest by just a bit, it then has a better water vapor supply and grows faster still. Soon it leaves the rest behind. When the humidity is very high, this plays out repeatedly, and the result can end up looking like a cluster of thin needles.

Bullet-shaped crystals are another variant of columnar growth. They form in clusters, with several bullets emanating from a single center. A cluster begins when a cloud droplet freezes into a jumble of several crystals, called a *polycrystal*. The axes of the different crystals in a polycrystal generally point along different directions, so several columnar crystals emerge from the polycrystalline core. As it grows out, the assembly becomes a cluster of bullet-shaped columnar crystals. The bullet clusters often come apart to yield isolated bullet crystals.

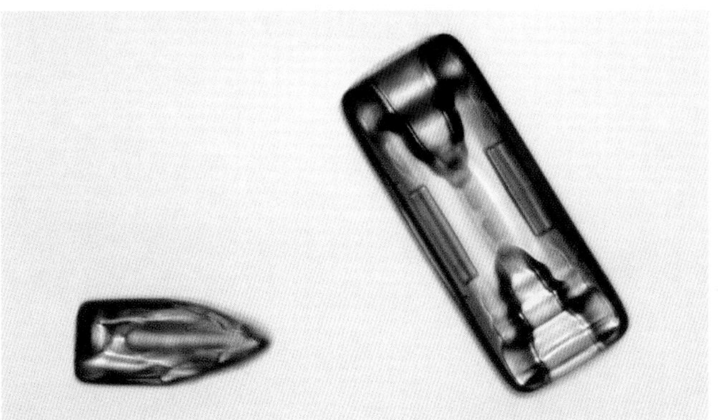

BULLETS

Above: Columns frequently form in clusters like these. Broken clusters yield bullet-shaped crystals, which can be readily found in many snowfalls.

SHEATHS

Left: Hollow columns can develop into hollow, sheath-like forms.

CAPPED COLUMN
These crystals started out as stubby columns and then switched to plate-like growth.

ASYMMETRIC CAPS
The end plates on capped columns frequently grow at different rates.

Capped Columns

Thin, flat plates and long, slender columns are the two extreme cases of snow-crystal growth. They each occur within a fairly narrow temperature range—the longest needles form near -5° C (23° F) and the thinnest plates form near -15° C (5° F). At intermediate temperatures, crystals grow into blockier forms. Since temperatures in the atmosphere can vary dramatically from place to place, a snow crystal may change its morphology substantially while it grows.

Often a snow crystal begins growing into a stubby column at intermediate temperatures and is subsequently carried by the wind to a region of the cloud at -15° C (5° F), where plates develop. In such cases, plates begin growing at both ends of the crystal, and we end up with a stout column capped by two thin plates. Nakaya called these *tsuzumi crystals*, after the Japanese drum with a similar shape. The more common name is *capped columns*.

Capped columns come in various shapes and sizes. The initial column can be long and thin or short and stubby, and the plates in turn can range from thin to thick. It all depends on the growth history of each crystal.

If the initial column is long enough, then two nearly independent plates can grow on its two ends, giving a symmetrical capped column. Often, however, the plates interfere with each other's growth, particularly on a short column. Then we run into yet another manifestation of the branching instability, resulting in a column capped with plates of unequal sizes.

If a capped column starts with two identical plates, then for a while they both grow at the same rate because they are growing in the same conditions. But if one of the plates grows just a bit larger than the other, the larger plate will stick out farther and find a better water supply, which causes it to grow faster. Soon it sticks out even farther, and thus it grows faster still. So the cycle goes until one plate wins out over the other. What results is an asymmetric capped column. This competition is inevitable with capped columns, although a given crystal may not have enough time to develop a large asymmetry between its plates.

COMPETING PLATES

Above: A surprising number of stellar snow crystals are actually capped columns with one dominant plate. Here the camera is focused on the smaller of the two plates.

CAPPED BULLETS

Left: These bullet crystals began growing plate-like caps.

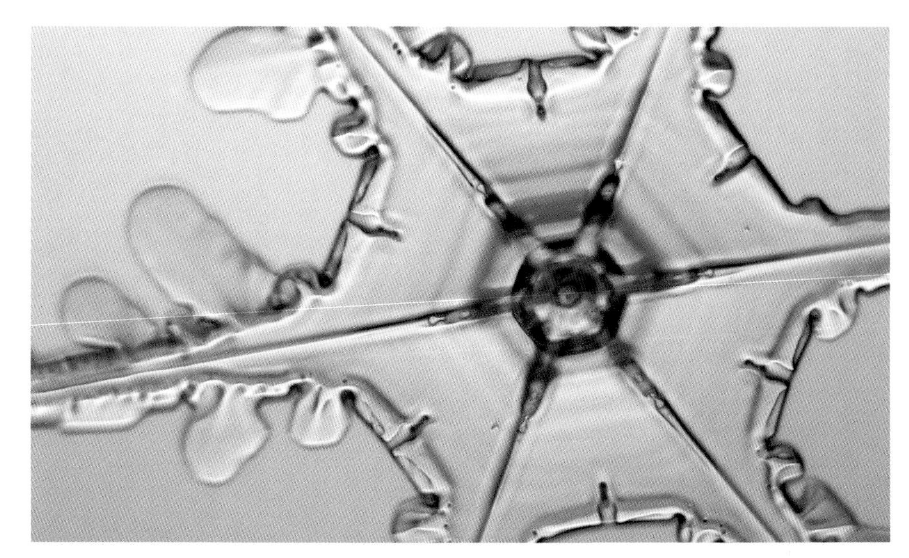

CENTRAL STRUCTURE
Here we see the central region of a capped column that became a large stellar crystal. The focus was on the large plate (top), between the plates (middle), and on the small plate (bottom).

Split Stars and Split Plates

The most dramatic example of a capped column with competing plates occurs when the plates develop dendritic branches. Diffusion causes the branches of one star to compete fiercely with the corresponding branches of the other, in yet another manifestation of the branching instability.

The competition between branches can produce a *split star* crystal. It starts on the ends of a small column, where two stellar crystals begin growing and branching. The scene soon becomes one of six growth races—each of the six arms of one stellar crystal compete with the corresponding arms of the other. Eventually the six races produce six separate outcomes. If the winning arms are from different ends of the column, then we have a split star.

Many of these bizarre morphologies, such as asymmetric capped columns or split stars, are not all that unusual. If you look at a typical snowfall you will probably see a substantial number of these forms. These special types reveal a great deal about how snow crystals develop.

Split stars and plates occasionally appear in a puzzling configuration in which the two competing plates do not point back toward a common center. The two separate plates may each contain three arms, and the arms of each plate may point to a well-defined center, but the two centers do not coincide.

Why these odd crystals look like this is a mystery. My best guess is that each probably started out as capped column that turned into a split star. Then at some point during the growth, the strain on the axle became too great and the plates came loose. But an instant after the plates broke apart, they collided and stuck together again. One plate moved relative to the other during this process, so in the end the two partial stars are not quite centered relative to one another.

This configuration is what I call a snowflake puzzler. It's a real snowflake, but it's not obvious what chain of events produced it. I enjoy speculating about such puzzles, and if you're so inclined, I invite you to try your hand at it too. If you get out your magnifying glass and look at some real snowflakes, you'll find there is no shortage of good puzzlers out there waiting to be solved.

STELLAR CONSTRUCTION

This crystal began as a small capped column—two plates joined by a thin column. As the plates competed for water vapor, the arms of the plates raced for dominance. Two arms from one plate and four from the other were the eventual victors. These winning arms grew into a stellar dendrite, while the defeated arms were left behind. The crystal broke apart when it hit the collection board.

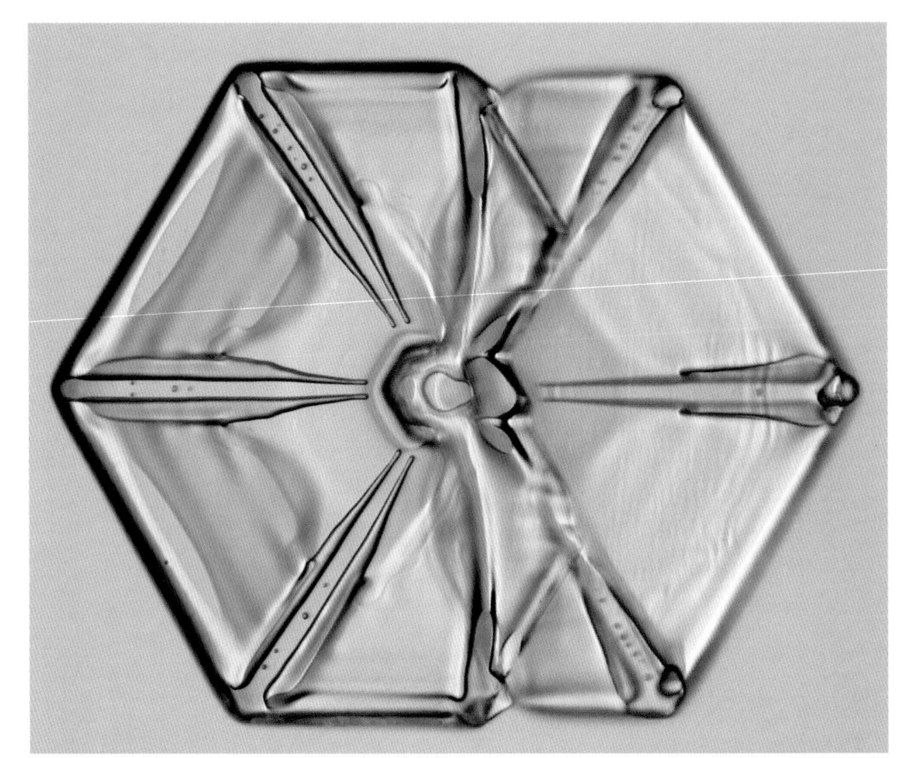

Split Plates

If the growth does not produce branching to make a stellar crystal, a snow crystal can end up with split plates.

Off-Center Stars

This split star has two plate centers that do not coincide. The crystal probably formed when the central axle of a split star became too weak, so the two plates flopped together.

Snowflake Puzzlers

Some split-plate crystals are so odd-looking they almost defy explanation.

Why is Snow White?

WHITE SNOW

Snow and hoarfrost cover an old barn and the surrounding landscape. (Photograph © Willard Clay)

Snow is made of small ice crystals, and a close inspection reveals that the individual crystals are clear, not white. However, when light travels from air to ice, or vice versa, some light is reflected, as when light reflects from a pane of glass. Since there are a lot of air and ice surfaces in a bank of snow, light shining into the snow reflects many times. After bouncing around a while inside the snowbank, some of the light bounces back out, and that is the light we see. Since all colors are reflected nearly equally well, the snowbank appears white.

At this point you might be asking, "If sunlight reflects off the snowbank, and sunlight is yellow, why doesn't the snowbank then appear yellow instead of white?" This question is deep and interesting.

The answer is that your brain somehow compares the light going in with the light coming out and figures out when something is white. For example, if you look at a white index card under various lights, you will always see that the card is white. Your brain somehow compensates for the color of the light source—although no one knows exactly how this works.

Twinned Crystals

Another lesson in snow-crystal hieroglyphics is about crystal twins. Twinning is a crystallography term referring to two separate crystals joined together in some precisely incorrect fashion. Compare it to buttoning up your shirt wrong: you make just one mistake in the beginning by putting the first button in the wrong buttonhole, and then you button up as usual and find the whole shirt is askew. Twinning works the same way; there's one molecular defect in the beginning, and then the whole crystal goes together wrong.

There are several ways ice crystals can get buttoned-up wrong, and one way makes a pair of twin columns. The defect starts in the crystal's nucleus, producing what looks like an ordinary columnar crystal. But a columnar twin is two crystals, one with a 60-degree twist relative to the other. The 60-degree twist means that the facets on one side of the prism are positioned exactly the same as the facets on the other side, so the twins appear as a normal column when they grow. But if you let the column evaporate a bit, it shows its true nature. The molecular bonds connecting the two crystals are weaker than the bonds in either of the twins. Thus the molecules at the join are more likely to evaporate. After some evaporation, the column displays a characteristic *evaporation groove* around its middle. You have to look for it, but it's common in columnar snowflakes.

Another example of twinning produces diamond-shaped forms that often grow alongside columnar crystals. The facet angles of the diamonds are not right for the usual hexagonal symmetry of the ice crystal, but a close look often shows an evaporation groove bisecting the form, indicating another example of twinning.

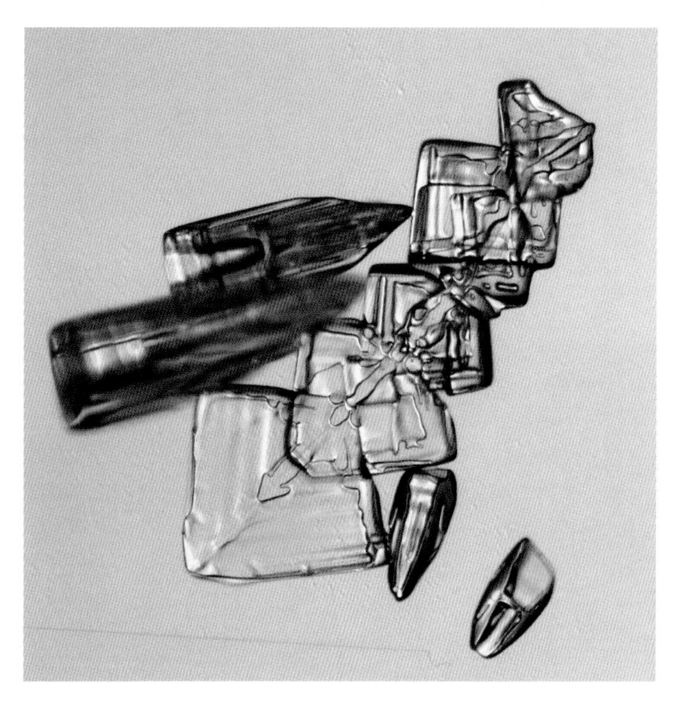

ARROWHEAD CRYSTALS
These unusual diamond-shaped plates fall along with columnar crystals, and the evaporation grooves show they result from crystal twinning.

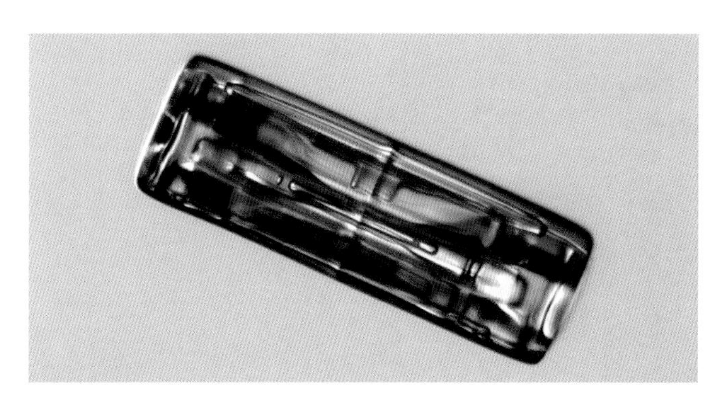

EVAPORATION GROOVES
A columnar crystal sometimes contains an evaporation groove that runs around the middle of the column. The presence of such a feature reveals that the two halves are crystal twins, incorrectly joined together at the molecular level.

Twelve-Sided Snowflakes and Double Stars

Twelve-sided snowflakes are one of nature's fascinating anomalies. Occasionally the process of turning a cloud droplet into a nascent snow crystal produces an unusual columnar twin consisting of two ordinary columns connected with a 30-degree twist, a variation of the 60-degree twinned column.

When a 30-degree twin column moves to -15° C (5° F), it grows two end plates that each follow the crystalline structure of their respective underlying columns, resulting in two separate six-branched plates rotated relative to each other. If the branches of the two plates are narrow, then after they grow out a bit, the tips are separated from one another. The separated branches do not compete with each other as vigorously as do the branches of a normal capped column. All twelve can then grow in a relatively normal fashion, creating a double star. Sometimes these look like symmetrical twelve-sided snowflakes.

I call this a snowflake, and not a snow crystal, because the whole structure is made of not one distinct crystal, but two. Although they are not common, twelve-sided snowflakes can be found if you keep your eyes open. Their abundance is not random; some storms produce a fair number of twelve-sided specimens, whereas other storms produce almost none. No one knows why.

DOUBLE STARS

How did this pair of stellar crystals form? It's unlikely they collided in midair, since both exhibit rare sectored-plate extensions. The pair probably started out as a twelve-sided snowflake in which the two stars were connected by a thin axle, as in a normal capped column. The 30-degree twist left room for both stars to develop independently, without a strong competition between the arms. Then at some point the axle broke and the assembly collapsed into the configuration you see here.

TWELVE-SIDED SNOWFLAKE

Twelve-sided snowflakes are made up of two six-branched crystals joined in the middle with a 30-degree twist. You won't see them every day, but they're not as hard to spot as you might think. Sometimes the centers of the two stars do not exactly coincide, as in this example.

Quadruple-Decker

Twelve-sided snowflakes, and even eighteen-sided snowflakes, are not the limit when it comes to multi-branched constructions. The specimen above exhibits no fewer than twenty-four branches. The branches are not symmetrically arranged, and some are a bit stubby, but clearly this is quite an unusual creation.

Eighteen-Sided Star

Above: Can you find all eighteen branches on this peculiar snowflake?

Multiple Capped Columns

Left: Multiple stars start out as columnar forms with numerous plate-like appendages. It's not clear how such structures develop.

Chandelier Crystals

In many snow crystals we see columnar growth turning into plate growth, producing a small menagerie of capped-column forms. It is also possible for plate growth to switch to columnar growth if the conditions dictate. I like to call these *chandelier crystals*.

STELLAR CRYSTAL WITH COLUMNS

Above: Although they are far less abundant than capped columns, you can sometimes find a plate-like stellar crystal that switched to columnar growth. The result in this case is a group of connected columns—a *chandelier crystal*.

HOMEMADE CHANDELIERS

Left, both photos: Although chandelier crystals are rare in the wild, we can make them pretty easily in the lab. The two pictures above show a crystal of this type, growing on the end of a thin ice-needle. First, a plate-like stellar dendrite was grown at -15° C (5° F), and then the crystal was moved to -7° C (19° F) for additional growth. Hollow columns subsequently formed at the ends of the dendrite arms. Just for fun, the crystal was then moved back to -15° C (5° F), where dendritic arms grew from each of the columns. I think this makes a rather fancy ice chandelier. To my knowledge, no natural snow crystal with such an odd morphology has ever been found; the temperature history would have to be quite unusual to create such a specimen in nature. (Photographs © Kenneth Libbrecht)

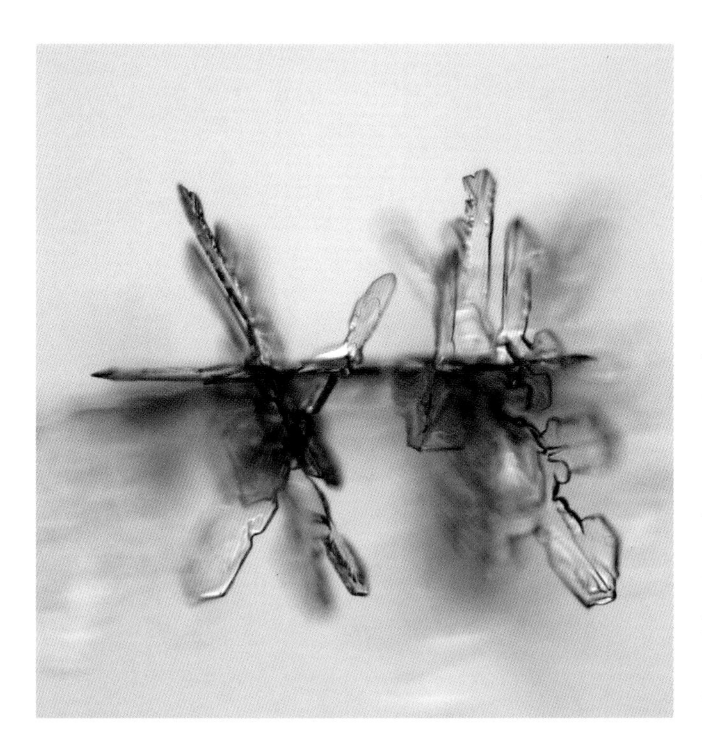

Spatial Dendrites

Many of the unusual snowflakes you find are *polycrystalline*—made of two or more distinct single crystal pieces. These often form when the original ice nucleus does not freeze into a single crystal, but rather into several crystals lumped together. The different crystal grains need not lock together with a well-defined relative orientation, and the subsequent growth need not be symmetrical. When a polycrystalline nucleus finds itself at -15° C (5° F), where dendritic branches readily grow, the result is a random collection of unrelated branches called a *spatial dendrite*.

Many people believe that all snow crystals possess fantastic six-fold-symmetric patterned forms. I assure you this is not the case. Imperfect snowflakes of various kinds, including asymmetric and otherwise malformed crystals, are far more common.

CLUSTERS OF BRANCHES
Above: This irregularly shaped snowflake grew from a nucleus containing many randomly oriented crystals. Branches grew out from the different crystal grains, each in a different direction.

THREE-DIMENSIONAL SNOW
Above: Spatial dendrites are common, and some snowfalls produce mostly these and other imperfect snowflake types.

Triangular Crystals

Occasionally one finds snow crystals that exhibit a distinctly triangular symmetry. It's not clear how these triangular crystals develop this symmetry, although it certainly has nothing to do with changes in the underlying symmetry of the ice lattice. The crystal maintains the usual hexagonal structure, as evidenced by the fact that the faceting on triangular crystals is no different than that on hexagonal crystals.

One scenario for making triangular crystals starts with a capped column that then makes a split star. If each star has three dominant branches, and these alternate between stars, then the split star will be like two three-armed crystals joined together. If the two stars break apart, you can be left with a pair of triangular crystals.

This split-star scenario likely plays out from time to time, and it probably explains some of the larger, elaborate triangular crystals. But many triangular crystals are too small to be thus explained, and in addition, we frequently observe smaller triangular crystals in the lab. It appears that some snow crystals simply decide early on to grow with three-fold symmetry.

These plate crystals develop with varying degrees of triangularity, forming most commonly around -2° C (28° F). The initial move to triangularity is probably accidental, as there is always some variation in the growth rates of the different sides of a crystal. The branching instability accentuates the process, so the triangular shape persists once it begins.

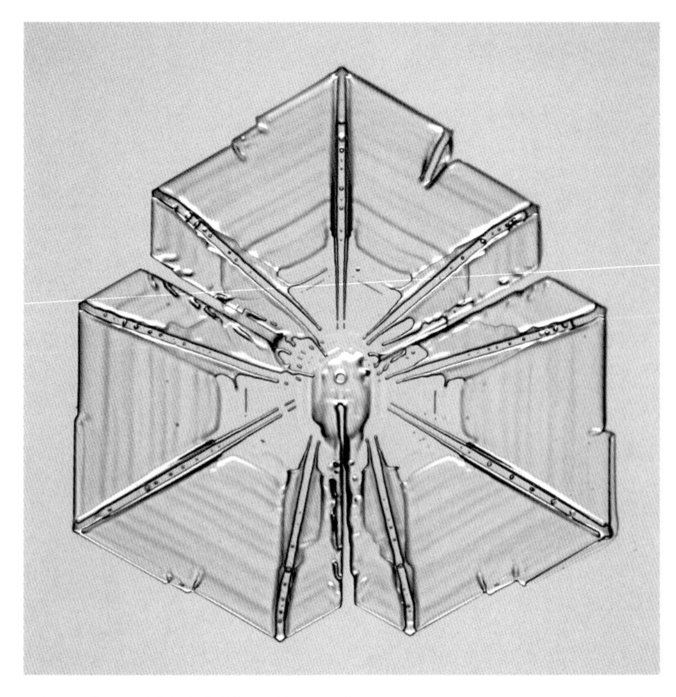

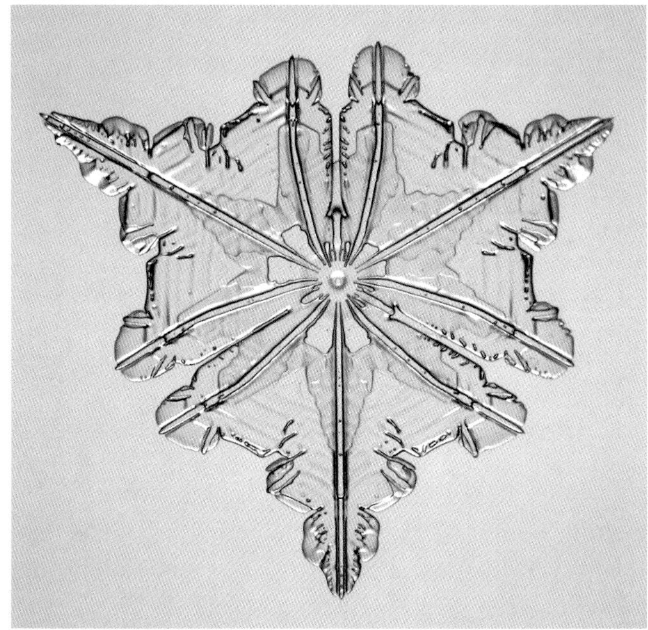

THREE BRANCHES

Above: Snow crystals with three roughly equivalent branches are unusual, and nice specimens tend to be small. These two each span about 0.6 mm (0.02 inches).

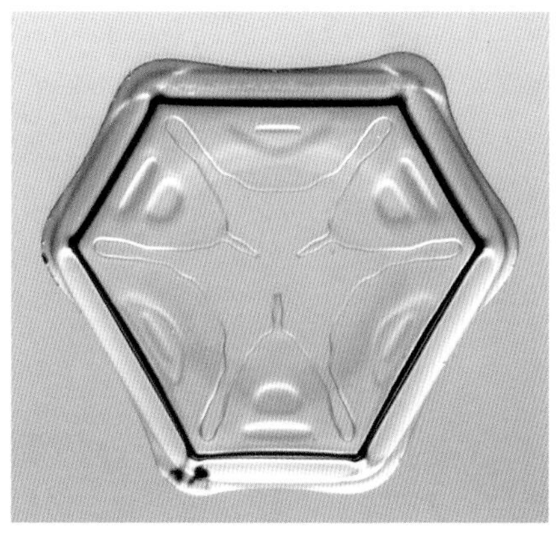

Half-Star

Above: This triangular crystal appears to be one side of a broken split-star crystal.

Accidentally Triangular

Above: The different branches of an ordinary stellar crystal don't always grow symmetrically. When three happen to lag behind by chance, a slightly triangular morphology results, as here.

Truncated Triangle

Right: These forms are nearly always small. The crystal is only 0.3 mm (0.01 inches) on a side.

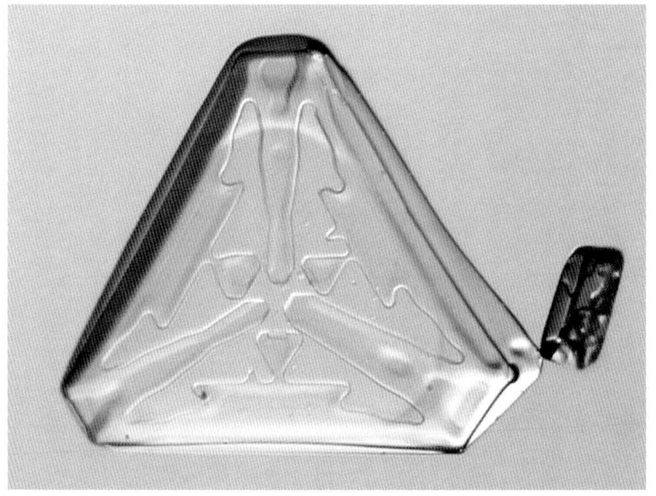

Hoarfrost

Surface Hoar

The most common type of hoarfrost forms overnight on the surfaces of snowbanks. When the sun goes down, the air temperature drops and the air in turn cools the surface of the snow. If the previous day was warm and the night cold, the snow surface-temperature can become lower than that inside the snowbank. When this happens, the temperature gradient causes water to evaporate from below and crystallize on the surface. The condensing water-vapor forms ice that bears a close resemblance to snow crystals, since the underlying phenomenon is the same. The crystals quickly melt again once the sun rises, so early morning is the best time to find nice surface-hoar crystals. (Photograph © Rachel Wing)

Ice crystals similar to snow crystals sometimes form on the ground as well in the sky. If supersaturated air is present at ground level, ice precipitates onto all convenient surfaces, resulting in *hoarfrost.* Since ice crystals growing from water vapor all follow the same growth rules, hoarfrost crystals are similar in appearance to snow crystals. Supersaturated air doesn't normally form near the ground, so hoarfrost crystals are not common. But they are around if you know where to look.

Caves and other enclosed structures sometimes house hoarfrost formations, provided there is a source of water vapor. If the air is still, hoarfrost can build up over days and weeks even if the supersaturation level is low. The most dramatic hoarfrost formations occur in the famed ice caves in Antarctica, where single crystals nearly a meter in size have been reported. These formations likely built up for years in the ever-frozen Antarctic.

One of my favorite hoarfrost formations was found by my wife one January, under the seat cover in an abandoned outhouse in North Dakota. All the ingredients for good hoarfrost were there—a cold, undisturbed site, liquid water at the bottom of what was basically a shallow well, and some weak convection to bring water vapor up from the depths. The site produced some beautiful cup-shaped crystals, about as large as a fingertip. And me without my camera!

Irregular Snow

Although we like best to look at the beautifully symmetrical crystals, the vast majority of crystals found in an average snowfall are irregularly shaped. A lot can happen to a snowflake during its travels through the clouds, and only a few snowflakes make it to the ground with near-perfect form.

Rimed Snowflakes

Since snow crystals grow in clouds, you might have wondered how they avoid running into all the small water droplets that make up clouds. The answer is that often they don't, and we can see the results on falling snowflakes.

When a droplet strikes an ice surface, it typically just sticks and freezes. The droplet becomes a small, round ice particle, and snow crystals frequently pick up a few such particles during their travels. These ice droplets are called *rime*.

Falling snow crystals are exposed to the full spectrum of rime coverage. The most common case is a light exposure, producing crystals with a few droplets here or there. If a droplet hits early, it can nucleate the growth of a new ice grain, resulting in a malformed polycrystalline snowflake. If the rime droplets hit late, you end up with a nice snow crystal decorated with just a few foreign droplets.

A snow crystal occasionally becomes so covered with rime droplets that it is barely recognizable as a snow crystal at all. When this happens, the resulting conglomeration of ice droplets is called a lump of soft hail, or *graupel*.

As with everything else about snow-crystal formation, the level of rime varies with time. In an hour or less, a snowstorm can change dramatically the quality of crystals produced. From heavy to light to no rime, from long, thin columns to large, flat plates, from simple prism crystals to complex stellar dendrites—it all depends on the ever-changing conditions in the clouds. A snowflake's character can be as fickle as the wind.

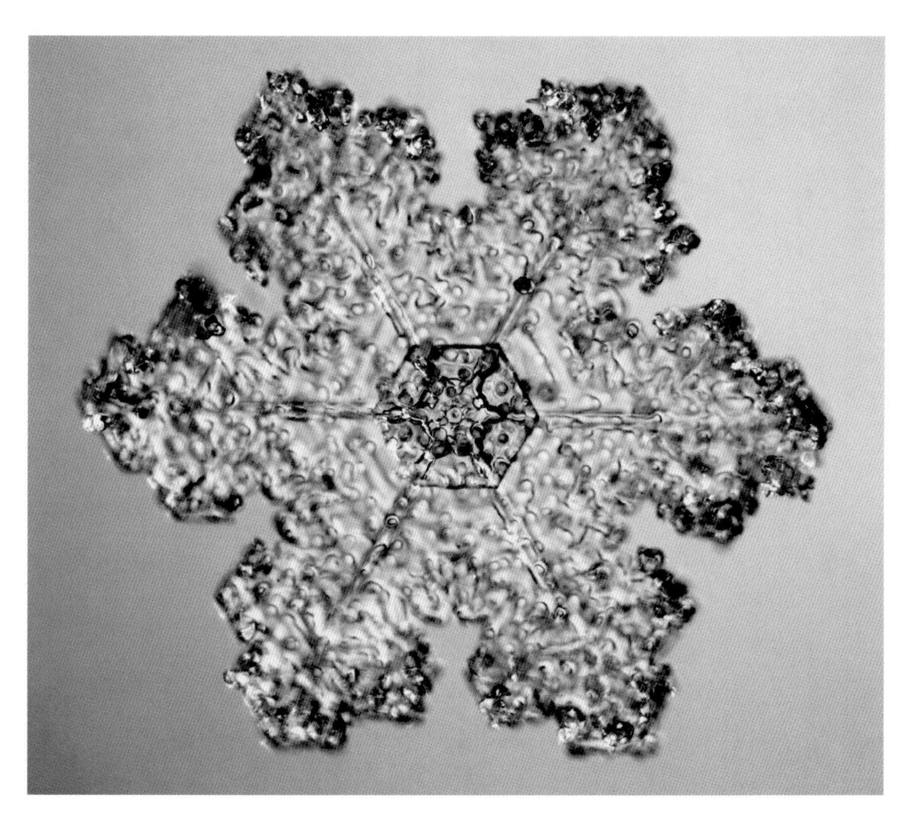

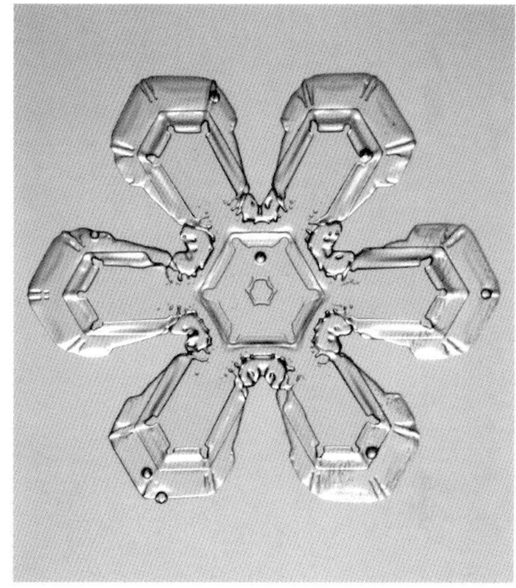

LIGHT RIME

Some snow crystals encounter just a few cloud droplets during their journeys.

HEAVY RIME

This crystal formed branches and nice six-fold symmetry in its early life. Then it ran into a dense cloud and was covered with rime droplets.

Chapter 8

In Search of Identical Snowflakes

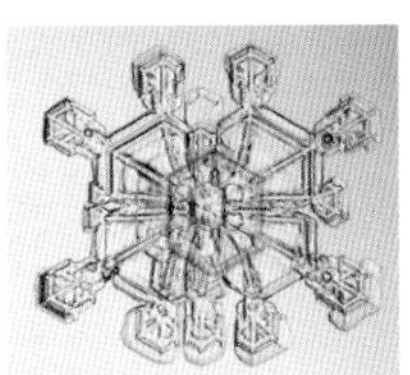

"The scientist does not study nature because it is useful; he studies it because he delights in it, and he delights in it because it is beautiful."
—*Jules Henri Poincare,* Science and Method, *1908*

Is it really true that no two snowflakes are alike? As the local snow "expert," I hear this often. It's a funny question, almost like a Zen *koan*—if two identical snowflakes fell, my inquisitive friend, who would know? And how can you say it is so, since you have not checked them all to find out?

Although there is indeed a certain level of unknowability to the question of snowflake alikeness, as a physicist I find that I can address this issue with some confidence. As I will demonstrate, the answer depends to a large degree on what you mean by the question. (Yes, physics does occasionally have its Zen-like qualities.)

SNOW STAR

Facing page: A snow crystal may contain a billion billion water molecules, give or take a few, and the chances of finding two snow crystals alike are nearly impossible.

First of all, you might take the meaning of "alike" to be exactly, precisely, alike. It is in fact a profound aspect of nature that there are some things that are exactly, precisely alike. For example, as far as we can tell in physics, all electrons are exactly, precisely alike. They have been so since the Big Bang, and they will be so into the indefinite future. The concept of "identical" is a cornerstone of quantum mechanics, and to say two electrons are identical means simply that there is no possible way to distinguish them. This is philosophical stuff, but nature can be like that. And regardless of whether we like it or not, nature seems to embrace the notion of identical objects.

Now a water molecule is considerably more complex than an electron, but if we restrict ourselves to water molecules made from two ordinary hydrogen atoms and one ordinary oxygen atom, then again physics tells us that all such water molecules are indistinguishable.

However, naturally occurring hydrogen comes in two different stable isotopes—hydrogen and deuterium. Ordinary hydrogen is made from an electron and a proton, whereas deuterium is made from an electron, a proton, and a neutron. On Earth, there is about one deuterium atom for every 5,000 ordinary hydrogen atoms. Hydrogen and deuterium are quite distinguishable atoms—we would just have to weigh them, since deuterium is about twice as heavy as ordinary hydrogen. Similarly, most oxygen atoms contain eight protons and eight neutrons, but on Earth about one in 500 has an extra pair of neutrons. And again the heavy isotope is certainly distinguishable from the normal variety.

Now, a snow crystal might contain a billion billion water molecules, give or take a few, and we see that on average about one in 500 of these will be different from the norm. Furthermore, these rogue atoms will be randomly scattered throughout the crystal, with many, many, many different possible configurations. Really, it's a mind-boggling number of possible configurations, more than all the protons and electrons in a million trillion universes like our own. Thus, the probability that two snow crystals would have exactly the same layout of these molecules is utterly, vanishingly small. It could snow day and night until the sun dies before two snow crystals would be exactly, precisely alike.

Even if we restrict ourselves to isotopically pure water molecules, made entirely from ordinary hydrogen and ordinary oxygen atoms, it's still unlikely that two snow crystals would be exactly, precisely alike. When a crystal grows, the molecules do not always stack together with perfect regularity. Crystals contain impurity atoms, stacking faults, and other types of defects. Mistakes happen. A typical snow crystal contains a huge number of these mistakes, which again are scattered throughout the crystal in a random fashion. One can again argue that the probability of two crystals growing with exactly the same pattern of molecular imperfections is utterly, vanishingly small.

Of course we assumed in all of this that we were looking at average-sized snow crystals. If a snow crystal contained only a handful of molecules—say, just ten—we might expect that every so often all ten molecules would be of the ordinary variety and all ten would stack

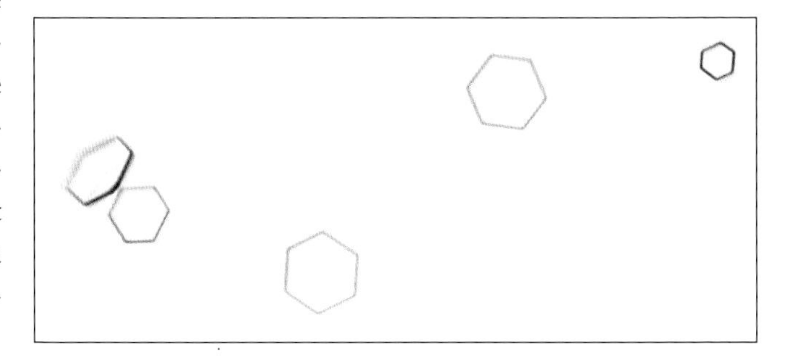

IDENTICAL SNOWFLAKES?

You might agree that the two snow crystals in the center of this picture look alike—or pretty close anyway. They were grown inside an artificial cloud in my laboratory, and they are only 0.03 mm (0.001 inches) in diameter, roughly half the diameter of a human hair. The trick to making two alike in this case was to make crystals by the thousands until two similar ones fell together by chance. (Photograph © Kenneth Libbrecht)

together without mistakes. Then yes, there's a reasonable probability that a pair of ten-molecule snowflakes would be exactly, precisely alike. It's up to you if you want to call a ten-molecule cluster of ice a snow crystal.

All right, so enough with demanding mathematical exactness. How about if we relax our definition of "alike" and say that two snow crystals are alike if they just look alike.

In a good optical microscope, the smallest features one can see are about 0.001 mm in size, which is about 3,000 times larger than a water molecule. Thus, we won't see any molecular-scale imperfections or isotopic differences. In this case, the simplest crystals will look pretty similar. Tiny snow crystals are usually fairly simple hexagonal prisms, and if you sifted through a collection of these you would soon find two that were essentially indistinguishable in a microscope. Since tiny simple crystals are fairly common in the atmosphere, it's fair to say that there are a great many snow crystals that look pretty much alike.

But that's only for simple crystals. What makes snow crystals so fascinating is that they form in intricate and beautifully symmetric shapes. For these crystals, it becomes much less likely that two will look alike. But here again it's a matter of degree. The closer you examine a particular specimen, the more distinct features you can make out inside the crystal—the particular arrangement of its branches, ridges, and so on. With dozens or hundreds of distinct features, each in a particular location, it soon becomes extremely unlikely that two crystals will ever have all the same features in all the same places.

There is a good analogy with faces. From a distance, faces often look similar, and you may mistake a stranger on the street for a good friend. But as you get closer, you can see more facial features, and up close faces all look quite different. Even identical twins look different, as any parent of twins will tell you.

So the question of identical snowflakes comes down to your requirements for being a snowflake, and what you really mean by being identical.

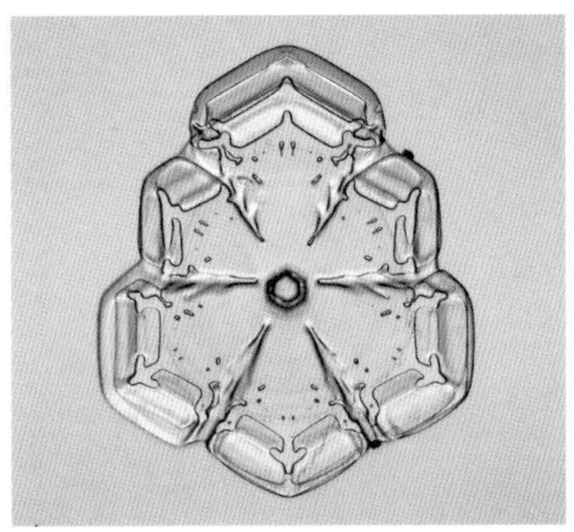

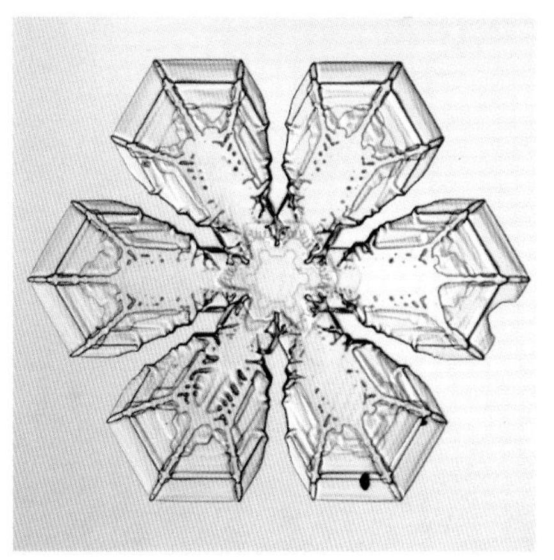

Thinking about Snowflakes

As a child watching the falling snow in North Dakota, I never imagined I would someday be thinking about the science of snowflakes. Now, having worked on the subject for many years, I still find it extremely rich and endlessly fascinating.

The physics governing snowflake growth touches on many topics, from the structure of crystals and their surfaces to the mathematical subtleties of morphogenesis and self-assembly. A careful examination of the inner workings of a snowflake reveals much more than just a sliver of ice. The symmetric patterns demonstrate the spontaneous generation of complex structures in the physical world.

I whiled away the cold winters of my youth throwing snowballs and building frozen fortresses out of packed snow. Now I construct designer snow crystals in my laboratory, trying to gain insights into the molecular dynamics of crystal growth. There are many mysteries left.

One thing I've learned from my research is that snowflakes are fascinating little structures that are full of surprises. It is my desire that this small book inspires you to look at snowflakes differently, to see them with new eyes. Perhaps the next time you find yourself surrounded by a gentle snowfall, you'll pick up a magnifying glass and discover firsthand the intriguing beauty of snowflakes. And should you find yourself examining one of these diminutive ice sculptures, I hope you will pause to think about what snowflakes really are, where they come from, and how they are created.

There is great beauty in a large, symmetrical stellar snow crystal. The beauty is enhanced by the magnifying lens that brings out the fine structures in the ice. The beauty is enhanced still further by an understanding of the processes that created it.

Afterword

Photographer's Statement

By Patricia Rasmussen

 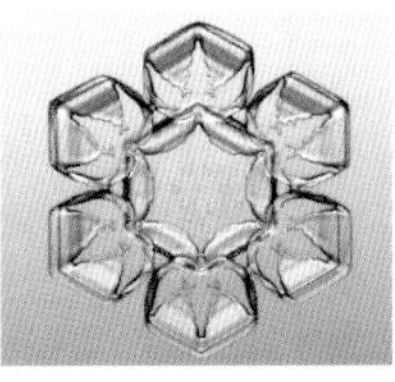 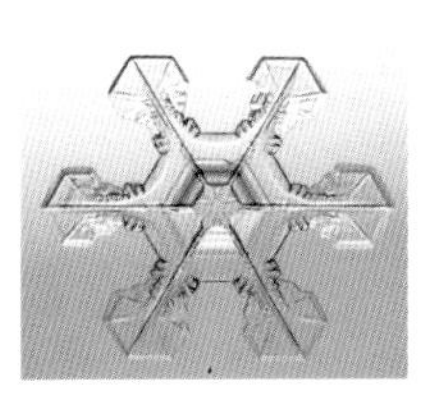

"If the sages ask thee why
This charm is wasted on the earth and sky,
Tell them, dear, that if eyes were made for seeing,
Then Beauty is its own excuse for Being."

—Ralph Waldo Emerson, "The Rhodora: On Being Asked, Whence Is the Flower," 1839

Late in 1997 I was introduced to Wilson Bentley's 1931 book of snow-crystal photographs. I was literally spellbound. I pored over Bentley's images, and immediately sought to learn everything I could about the art of photographing snow crystals. How did Bentley create such detailed, beautiful photographs and amass such a large collection? His inspiring dedication and accomplishments are still nothing short of amazing to me. Bentley took his first successful snow-crystal photomicrograph on January 15, 1885. I took my first successful photograph of a snow crystal on January 22, 2000—almost 115 years later to the day.

EPIPHANY
Facing page: A small bit of wonder in the everyday world.

When I saw my first snow crystal through the viewfinder of my camera, I almost fell over backwards. It was an epiphany and more than took my breath away. No longer were snowflakes contrived facsimiles that adorned wrapping paper and greeting cards. No longer was snow something to complain about, flee from, shovel away, ski and play in, throw at human targets, or lose traction in. Even though I'd often watched snow falling and have stood in hushed reverence in a landscape masked and insulated with fresh fallen snow, it was through optical magnification that I had the opportunity to truly see snow.

"In nature, specialness is not for a privileged few. Every kind of life is unique, and it is these differences, these millions of differences that make living on earth the grand adventure that it is."
—*Les Blacklock,* Ain't Nature Grand!, *1980*

I suspect snow-crystal photography was both a passion and an addiction for Wilson Bentley. In a 1902 article for *Monthly Weather Review* magazine, Bentley wrote: "It is extremely improbable that anyone has as yet found, or, indeed, ever will find, the one preeminently beautiful and symmetrical snow crystal that nature has probably fashioned when in her most artistic mood." This is the fuel for obsession. It is so difficult to deny the anticipation of the next crystal's design. You never know—will the next snow crystal be the one that surpasses all others? Their individual differences enhance each other, and that's why viewing a collection of photographs is all the more fascinating. No single snow crystal has it all.

A snow-crystal photographer can often hardly muster the willpower to pack it up for the day—or night, for that matter—until the snow stops falling. I made this notation on January 19, 2002: "Minute plates before sunrise—so small you couldn't feel them or really see them save for the outdoor sodium vapor light shining through them like diamonds. As morning grew longer, so did little arms on the little plates." I heeded the moment by putting my woolen outer clothes on right over my pajamas so as to not waste any time. The air was cold, the glistening crystals were free from rime, and I was excited to get outdoors and photograph.

"Silent snowflakes softly fall,
Little geometric wonders
Tiny jewels holding the universe
In each unique design
Each one a masterpiece."
—*Diana Anderson, untitled, 2001*

For me, snow-crystal photography is an escape from the stress of modern life into a sanctum of beauty and wonder and solitude. It's a treasure hunt. It's a photographic harvest from nature's bounty. I do move quickly when I'm at work and don't have time to stop and savor the snowfall, however, because an isolated crystal lasts but a moment. It is only later when I get my pictures in front of me for editing that I can finally linger, get to know them, and appreciate the finer details of each design.

During the winter of 2001–2002, I had the privilege to use an apparatus designed and built by Ken Libbrecht for high-resolution snow-crystal photography. All of my photographs are of real, natural snowflakes. The images were not digital creations. They were, however, digitally edited and optimized through processes like cleaning dust and debris and adjusting color balance.

It was great fun to experiment with colored light to enhance an otherwise colorless crystal. I was especially obsessed with finding a technique that would render a multicolored crystal radiant and full of symbolism. I used software on my computer to design and print transparent light filters that manifested multiple colors within the crystal at the time of exposure. A few images show the light distributed as a color-wheel effect within the facets of the crystal. These color inclusions were not computer generated, and I consider them rare gifts.

On January 6, 1858, Henry David Thoreau wrote in his journal that a snowflake and a dewdrop are the product of enthusiasm. These images were photographed with enthusiasm. May you too find new wonder in a winter's snowfall.

"We die on the day when our lives cease to be illumined by the steady radiance, renewed daily, of a wonder, the source of which is beyond all reason."
—*Dag Hammarskjöld,* Markings, *1966*

Photographing Snowflakes

By Kenneth Libbrecht

Capturing the fine details of a snow crystal requires a good camera, high-quality microscope objectives, and rugged mounting hardware to keep things stable. The photo-microscope shown here was built by the author and used for most of the photographs in this book. Many of its unique features were designed specifically for snowflake photography. Cameras generally don't work in extreme cold, so this one is kept in a temperature-regulated enclosure (shown here with the top removed). The focus and other camera controls are such that they can be operated with gloved hands.

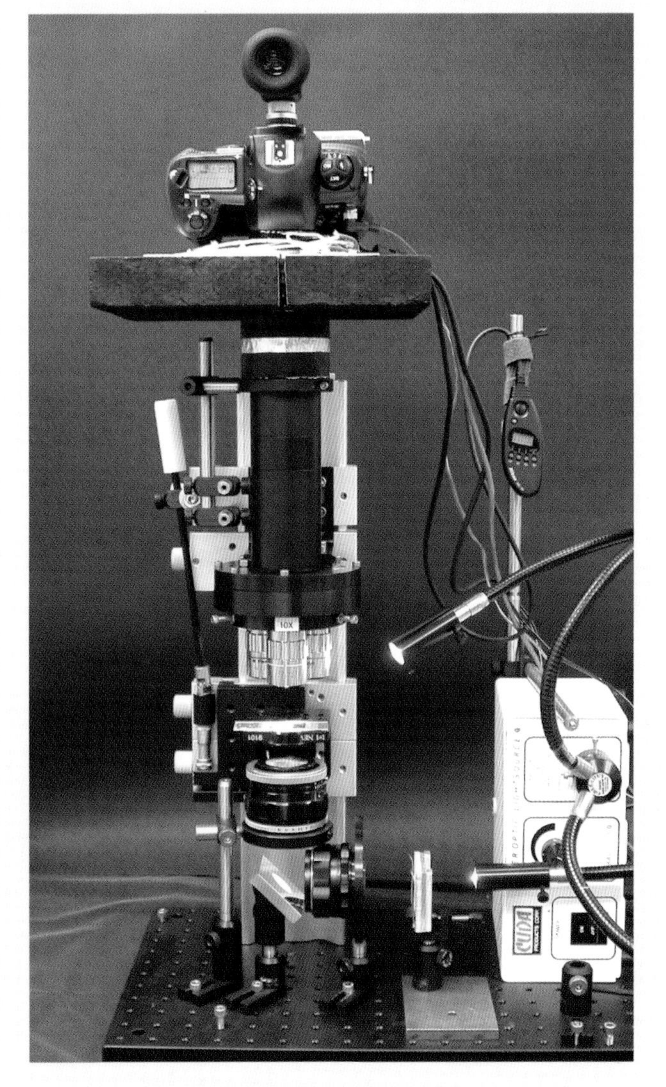

Since snowflakes are made of ice, which is both colorless and transparent, good lighting is especially important for snowflake photography. This microscope incorporates color filters to provide background effects and bright highlights. These add depth and help bring out the three-dimensional structure of the crystals.

Arranging snowflakes under the microscope requires patience and a gentle touch. One starts by examining snowflakes on a large collecting surface to find a suitable specimen. The chosen crystal can then be transferred to a glass microscope-slide using a soft feather or brush.

The entire photography procedure must be done in the cold. And it must be done quickly, before the minute patterns in a freshly fallen snowflake evaporate forever.

INDEX

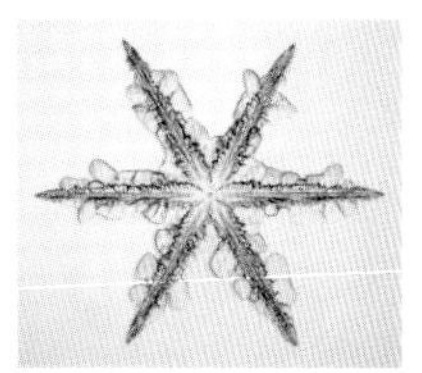  

atmospheric halos, *68–69*
Bentley, Wilson, *30, 32, 43, 63, 107, 108*
bullet snow crystals, *77*
capped columnar snow crystals (tsuzumi snow crystals), *78–80, 81, 91, 92, 94*
chandelier snow crystals, *92*
Chickering, Frances, *27, 28–29*
Cloud Crystals: A Snow-Flake Album (Chickering), *27, 28–29*
cloud seeding, *62*
columnar snow crystals, *20, 23, 25, 76, 85*
crystals, *38–41*
  definition, *38–40*
  formation, *40–41*
Descartes, René, *27–28, 39*
deuterium, *102*
diamond-dust snow crystals, *68–69*
double star snow crystals, *86*
evaporation groove, *85*
facets, *39–41*
  basal, *41*
  hexagonal prism, *41, 45, 52, 56, 68*
  prism, *41*
graupel, *99*
hail, *99*
hoarfrost, *97*
hollow columnar snow crystals, *76–77*
Hooke, Robert, *28*
Humphreys, W. J., *30*
hydrogen isotopes, *102*
ice nucleation, *44, 47, 62*
ice types, *39*
  Ice Ih, *36, 38, 39*
  Ice Ix, *39*
Kepler, Johannes, *35–36, 41, 43*
Les Météores (Descartes), *27–28, 39*
Micrographia (Hooke), *28*
morphogenesis, *51–52, 104*
morphology diagram, *45, 47, 57*
Nakaya, Ukichiro, *43–45, 47, 57, 63, 78*
needle snow crystals, *76, 77*
Pliny the Elder, *38, 39*
rime, *19, 99*
sectored plate snow crystals, *24*
self-assembly, *38–39, 40*
Six-Cornered Snowflake, The (Kepler), *35–36*
sleet, *21*
snow clouds, *60–61*
snow crystals,
  affects of humidity, *45–47*
  affects of temperature, *45–46*
  affects of weather, *47*
  artificial propagation, *44–47, 55*
  branching instability, *52–53, 54, 56*
  color, *84*
  complexity, *47, 52–53*
  coordinated arm growth, *47*
  definition, *18*
  dendrites, *54*
  diffusion-limited growth, *52, 54, 56*
  faceting, *47, 52–53, 56*
  formation, *21, 38, 47, 52*
  ice nucleation, *44, 47, 62–63*
  identical, *101–103*
  morphogenesis, *51–52, 104*
  morphological balance, *56*
  riming, *19*
  self-assembly, *38–39, 40*
  six-fold symmetry, *18–20, 35–36, 38, 52*
  symmetry, *18, 20, 23, 35–41, 47*
snow crystal types, *63*
  bullet, *77*
  capped columnar, *78–80, 81, 91, 92, 94*
  chandelier, *92*
  columnar, *20, 23, 25, 76, 85*
  diamond dust, *68–69*
  double star, *86*
  hollow columnar, *76–77*
  needle, *76, 77*
  sectored plate, *72–75*
  spatial dendrite, *93*
  split plate, *81–83*
  split star, *81–82, 94, 95*
  stellar dendrite, *18, 21, 69–71, 81, 92*
  triangular, *94–96*
  twelve-sided, *25, 86–90*
  twinned, *85, 86*
Snow Crystals (Bentley), *30*
snowflake, *see also* snow crystal
  definition, *18*
  formation, *18*
spatial dendrite snow crystals, *93*
split plate snow crystals, *81–83*
split star snow crystals, *81–82, 94, 95*
stellar dendrite snow crystals, *18, 21, 69–71, 81, 92*
sun dogs (22-degree parhelia), *68–69*
triangular snow crystals, *94–96*
twelve-sided snow crystals, *25, 86–90*
22-degree parhelia, *see* sun dogs
twinned snow crystals, *85, 86*
Vonnegut, Bernard, *62*
Vonnegut, Kurt Jr., *39, 62*

About the Author

Kenneth Libbrecht was raised in North Dakota and educated at Caltech and Princeton, receiving his Ph.D. in physics in 1984. He subsequently joined the faculty at Caltech, where he is currently professor of physics and chairman of the physics department. Dr. Libbrecht has published numerous articles on a range of scientific topics, including the free oscillations of the sun and stars, ultra-cold atomic gases, the detection of gravitational radiation, and the mechanics of crystal growth. He resides in Pasadena with his wife and two children. The latest news and views of snow-crystal research can be found at his website, *www.snowcrystals.net*.

About the Photographer

Patricia Rasmussen was born and raised in Wisconsin, where as Sinclair Lewis stated, "Snow isn't just a season, it's an occupation." Discovering Wilson Bentley's book *Snow Crystals* in 1997, she was inspired to investigate snowflake photography. Using camera equipment bought in eBay auctions, she took her first snow-crystal photos during the winter of 1999–2000. Seeing her first snowflake through the viewfinder, she almost fell over backward, the vision taking her breath away. Working with Kenneth Libbrecht's specially built snow-camera apparatus, she began creating the photos reprinted in this book in the winter of 2001–2002. Her photography has appeared in *USA Today*, *Wisconsin Trails*, and *The Tennessean*, and has been used in science and math curriculum, videos, company logos, student science fairs, embroidery designs, Christmas cards, and an avalanche-awareness school. Patricia acknowledges the support and encouragement of her husband, Eric, and two sons, Brian and Joel, which allowed her the freedom to photograph snowflakes.

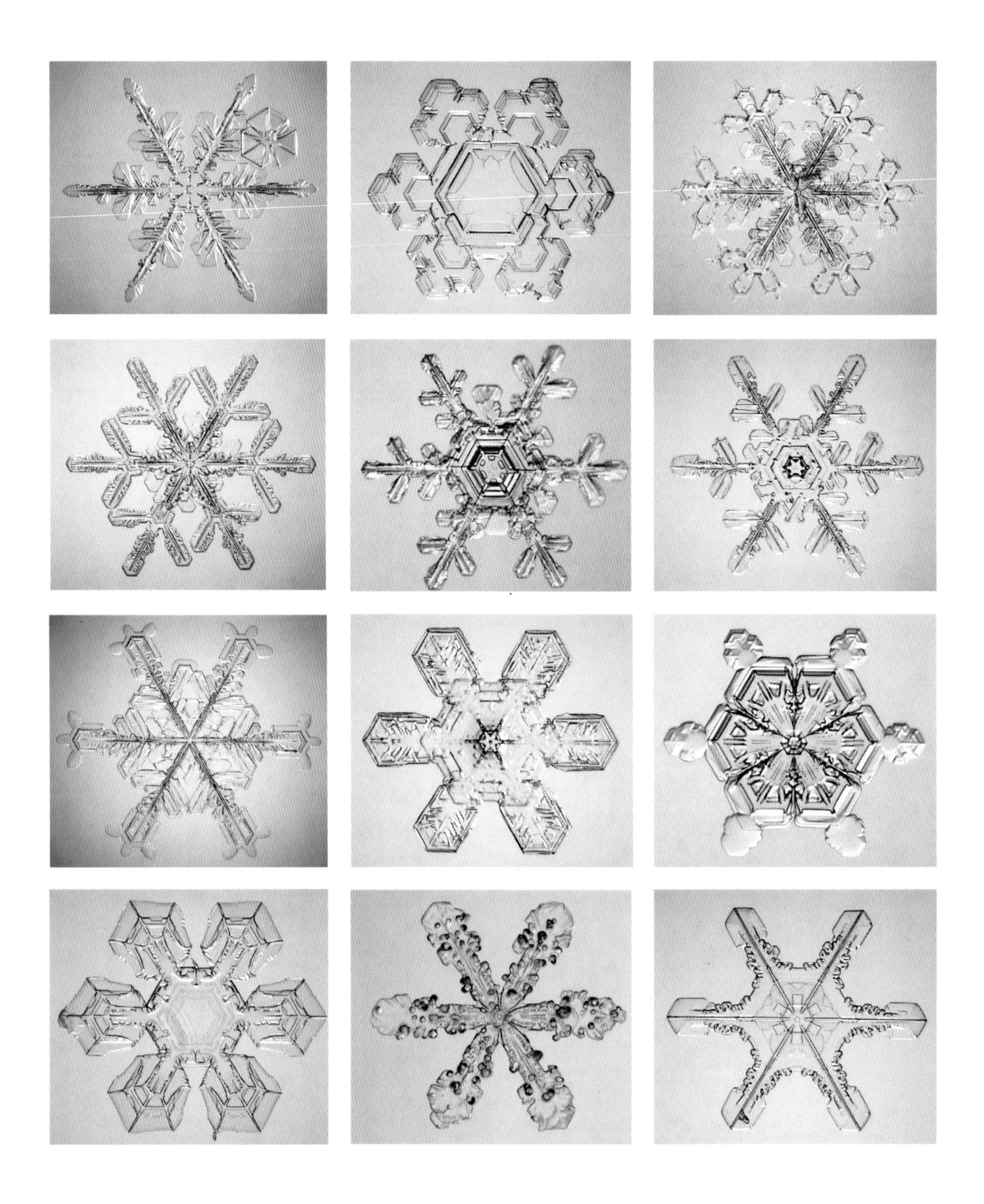